AF887340

Cells Tissues Organs

in vivo, in vitro

Founded 1945 as 'Acta Anatomica' by R. Chambers, G. Glimstedt, T. Peterfi and G. Wolf-Heidegger
Continued 1962–1974 by E.A. Boyden, 1955–1980 by A. Delmas, 1972–1980 by F. Walberg, 1945–1980 by G. Wolf-Heidegger, 1981–1988 by R. O'Rahilly, Davis, Calif., 1989–1990 by G.E. Goslow, Jr., Providence, R.I., 1981–1992 by W. Lierse, Hamburg

KARGER

Printed in Switzerland on acid-free paper by Reinhardt Druck, Basel

Appears 6-weekly: 2 volumes per year (8 issues)

Guidelines for Authors

Cells
Tissues
Organs

Aims and Scope
Cells Tissues Organs is an international forum for communication among scientists interested in current research on functional morphology and morphogenesis at all levels of organization, from subcellular to macroscopy, with emphasis on humans and higher vertebrates. The journal offers rapid exchange of novel and significant information gained from experimental and theoretical investigations on all aspects that are pertinent to the formation and transformation of biological structure.

Topics of special interest include
Developmental and reproductive biology, cell biology, tissue engineering, artificial organs, in vitro systems and transplantation. Papers highlighting the links between structural and functional aspects are preferred.

Types of Papers
The journal publishes: (1) Original papers on experimental research. (2) Theoretical contributions. These will appear in the section 'Hypotheses and Concepts' and may include provocative hypotheses, provided that they are innovative, well founded and show promise for new avenues of research. (3) Short reviews/perspectives. While (1) and (2) are submitted contributions, reviews/perspectives (3) are invited. Suggestions concerning reviews on current topics of high interest are welcome.
All manuscripts must be accompanied by a covering letter signed by all authors. Assurance should be given that the manuscript is not under simultaneous consideration by any other publication. Reported studies using animal and human subjects must have been approved by the appropriate institutional review boards.

Submission
Manuscripts from North and South America should be sent to:

Prof. A.W. English, PhD
Emory University School of Medicine
Dept. of Anatomy and Cell Biology
Atlanta, GA 30322 (USA)
Fax +1 404 727-3677
E-Mail cto@cellbio.emory.edu

Manuscripts from Japan and South-East Asia to:

Prof. Dr. Osamu Ohtani
Department of Anatomy II
Toyama Medical and Pharmaceutical University
2630 Sugitani
Toyama, 930-01 (Japan)
Fax +81 764 34 5010

Manuscripts from all other countries to:

Prof. H.-W. Denker
S. Karger AG
Editorial Office 'Cells Tissues Organs'
P.O. Box
CH–4009 Basel (Switzerland)
Tel. +41 61 306 13 57
Fax +41 61 306 14 34
E-Mail l. haas@karger.ch

Mailing address for courier deliveries only:

Allschwilerstr. 10
CH–4055 Basel (Switzerland)

All manuscripts are subject to peer review. Submission of an article for publication implies the transfer of the copyright from the author to the publisher upon acceptance. Accepted papers become the permanent property of Cells Tissues Organs and may not be reproduced by any means, in whole or in part, without the written consent of the publisher. It is the author's responsibility to obtain permission to reproduce illustrations, tables, etc. from other publications.

Format and Arrangement
Manuscripts should be written in English and be as concise as possible. They should be submitted in quadruplicate (with four sets of original illustrations), typewritten double-spaced on one side of the paper, with a wide margin. *Lines should be numbered on the margin.*

Title page: Title, the authors' names, and the institute where the work was conducted, a short title for use as running head, the full address of the author to whom correspondence should be sent. Please also supply phone and fax numbers, as well as your e-mail address. If different, indicate in addition a full address for revisions and proof reading separately.

Key words: For indexing purposes provide a list of up to 5 key words.

Abstract: Each paper needs an abstract of up to about 20 lines summarizing the essential new information communicated.

Subheadings: Introduction; Materials and Methods; Results; Discussion.

Small type: Paragraphs which can or must be set in smaller type (methods, etc.) should be indicated with a 'p' (petit) in the margin on the left-hand side.

List of Abbreviations: Provide a list of the abbreviations used in your paper. Please list them on a separate page of the manuscript. The list will be printed as a footnote at the bottom of the first page of your article.

Footnotes: With the exception of the list of abbreviations, footnotes should be avoided. When essential, they are numbered consecutively and typed at the foot of the appropriate page.

Nomenclature: The nomenclature used should conform, wherever possible, to the current edition of the Nomina Anatomica (Terminologia Anatomica), Nomina Anatomica Veterinaria, or to the Nomina Anatomica Avium.

Tables and illustrations: Tables and illustrations (both numbered in Arabic numerals) should be prepared on separate sheets. Tables require a heading and figures a legend, also prepared on a separate sheet. For the reproduction of illustrations, only good drawings and original photographs can be accepted; negatives or photocopies cannot be used. Due to technical reasons, figures with a screen background should not be submitted. When possible, group several illustrations on one block for reproduction (max. size 181 × 223 mm) or provide crop marks. On the back of each illustration, indicate its number, the author's name, and 'top' with a soft pencil.

Color illustrations: During the review process, the decision will be made as to which color illustrations of a manuscript can be reproduced free of charge. Authors will be informed of this at the time of acceptance of their paper. Otherwise, these illustrations will be reproduced at the author's expense. Up to 6 color illustrations per page can be integrated into the text at the reduced price of CHF 660.– / USD 545.00 per page.

References: References in the text should be made by author's name followed by the year of publication, arranged chronologically, then alphabetically. When there are more than two authors, use the first author's name and et al. When references are made to more than one paper by the same author, published in the same year, designate them as a, b, c, etc. The list of references should include only those publications which are cited in the text. Material submitted for publication but not yet accepted may be cited in the text as 'unpublished data' but must not be included in the reference list. In the list of references, arrange references alphabetically listing all authors, then year of publication. Abbreviate journal names according to Index Medicus.

Examples
(a) Papers published in periodicals: Sun, J., H. Koto, K.F. Chung (1997) Interaction of ozone and allergen challenges on bronchial responsiveness and inflammation in sensitised guinea pigs. Int Arch Allergy Immunol *112:* 191–195.
Hadler, W.A., S.R. Silveira (1981) Técnica histoquímica que permite discriminar o colesterol de substâncias afins e de seus ésteres. Cienc Cult *33(suppl):* 586–587.
(b) Monographs: Matthews, D.E., V.T. Farewell (1996) Using and Understanding Medical Statistics, ed 3, revised. Basel Karger.
(c) Edited books: Parren, P.W.H.I., D.R. Burton (1997) Antibodies against HIV-1 from phage display libraries: Mapping of an immune response and progress towards antiviral immunotherapy; in Capra, J.D. (ed): Antibody Engineering. Chem Immunol. Basel, Karger, vol 65, pp 18–56.

Further details: Authors may consult the 'Guide to Writing Articles in English' by R. O'Rahilly (Acta Anat *131*: 1–2, 1988).

Galley proofs
Unless indicated otherwise, galley proofs are sent to the first-named author and should be returned with the least possible delay. Alterations made in proofs, other than the correction of printer's errors, are charged to the author.

Reprints
The main authors of all papers appearing in the 1999 issues will receive a one-year subscription and 50 reprints free of charge. Order forms and a price list are sent with the galley proofs. Orders submitted after the issue is printed are subject to considerably higher prices.

KARGER

Fax + 41 61 306 12 34
E-Mail karger@karger.ch
www.karger.com

The Guidelines for Authors are available at:
www.karger.com/journals/cto/cto_gl.htm

General Information

Cells
Tissues
Organs

Publication data: 'Cells Tissues Organs' is published 8 times a year. Volumes 164–165 with 4 issues each appear in 1999.

Subscription rates: Subscriptions run for a full calendar year. Prices are given per volume.
Personal subscription:
Subscription + postage and handling
CHF 97.– + CHF 15.70 Europe
CHF 97.– + CHF 23.70 Overseas
USD 74.00 + USD 15.70
(Must be in the name of, billed to, and paid by an individual. Order must be marked 'personal subscription'.)
Institutional subscription:
Subscription + postage and handling
CHF 644.– + CHF 19.60 Europe
CHF 644.– + CHF 29.60 Overseas
USD 492.00 + USD 19.60
Airmail postage: CHF 24.–, USD 20.00.
(Extra per volume.)

Microforms: Please direct all enquiries and orders to: University Microfilms Inc.
300 North Zeeb Road, P.O. Box 1346
Ann Arbor, MI 48106-1346 (USA)

Subscription orders: Orders can be placed at agencies, bookstores, directly with the Publisher
S. Karger AG, P.O. Box,
CH–4009 Basel (Switzerland)
or with any of the following distributors:
Deutschland: S. Karger GmbH, Postfach
D–79095 Freiburg
France: Librairie Luginbühl
36, bd de Latour-Maubourg, F–75007 Paris
Great Britain, Ireland: S. Karger Publishers
58 Grove Hill Road
Tunbridge Wells, Kent TN1 1SP
USA: S. Karger Publishers, Inc.
26 West Avon Road, P.O. Box 529
Farmington, CT 06085
India, Bangladesh, Sri Lanka:
S. Karger India Subscription Agency
B–5/132 Safdarjung Enclave
New Delhi-110029
Japan: Katakura Libri, Inc.
2F, Marushima Bldg, 4-6-17, Yushima,
Bunkyo-ku, Tokyo 113-0034

Change of address: Both old and new address should be sent to the subscription source.

Single issues and back volumes: Information can be obtained through the Publisher.

Advertising: Correspondence and rate requests should be addressed to the Publisher.

Bibliographic indices: This journal is regularly listed in bibliographic services, including *Current Contents®*.

Copying: This journal has been registered with the Copyright Clearance Center (CCC), as indicated by the code appearing on the first page of each article.
For readers in the US, this code signals consent for copying of articles for personal or internal use, or for the personal or internal use of specific clients, provided that the stated fee is paid per copy directly to
Copyright Clearance Center Inc.
222 Rosewood Drive
Danvers, MA 01923 (USA)
A copy of the first page of the article must accompany payment. Consent does not extend to copying for general distribution, for promotion, for creating new works, or for resale. In these cases, specific written permission must be obtained from the copyright owner,
S. Karger AG,
P.O. Box, CH–4009 Basel (Switzerland).

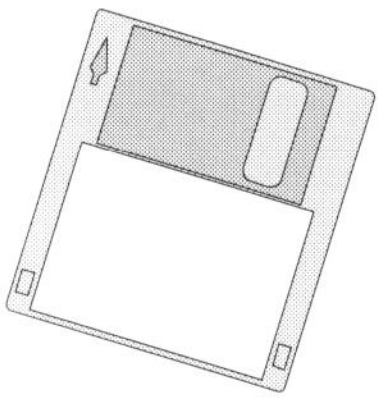

Disk Submission

A **comprehensive version** of these guidelines for the **submission of manuscripts and illustrations on disk** can be accessed online under **www.karger.com/journals/disksub.htm** or requested as a leaflet from the editorial department.

Please follow the general rules on style and arrangement in the **"Guidelines for Authors"** of the journal. Submit a **double-spaced printout** of your manuscript matching the file on disk exactly.

Technical Information

Store only data pertinent to this manuscript. Include your fonts (screen and printer fonts for Macintosh). For a **text only** manuscript, use a 3.5 inch disk. For **text and illustrations,** use 3.5 or 5.25 Magneto Optical Disc; CD-ROM (DOS or hybrid); iomage Zip/DOS (100 MB), or 44 MB, 88 MB or 200 MB SyQuest. Compress extensive data using ZIP format.

Data Description

Label your disk with the **title of the journal or book** for which it is submitted; a **short version** of the paper's title; your **name,** postal and e-mail address, fax and phone number; **hard and software** used, including the version; and the **complete list of the file names.** File names consist of eight characters before the dot identifying the file and three characters after the dot identifying the application.

Text and Format

The **preferred word-processing package is MS-Word.** Other commonly used PC text programs and ASCII formats are accepted.
Update your data when you revise subsequent to the reviewer's or editor's decision and mark revisions clearly. Manuscripts accepted for publication are subject to **copy editing:** changes and/or questions will be marked in the manuscript, which will be returned to you with your disk and the galley proofs for checking.
Enter your text continuously flush left. Do not use any layout functions. Do not split words at the end of a line. **Do not use indentations.** Use a blank line before and a hard return at the end of paragraphs. **Arrange headings flush left.** Insert only **one space after punctuation marks.** Insert Greek letters, mathematical symbols etc. with your word-processing program. Use **boldface and italics** as well as sub- and superscript where appropriate.
Tables are part of the text and should be placed at the end of the text file. Separate the individual parts of a table by lines. Use the tabulator instead of spaces. Format operation data on one line in the same column. **Legends to illustrations are part of the text** and should also be listed at the end of the text file.

Illustrations

Store illustrations as separate files: do not embed them in your text. Export **line drawings/vector graphics** as TIF, EPS or WMF format. Use Photoshop for processing and retouching **scanned halftone images.** Save the original scan and the processed version. Export **black and white or color images** as TIF or EPS format in their anticipated size in print. **For detailed information** on how to submit illustration data on disk please consult the leaflet or **www.karger.com/journals/disksub.htm.**

KARGER
Fax+ 41 61 306 12 34
E-Mail karger@karger.ch
www.karger.com

The Journal Home Page is available at:
www.karger.com/journals/cto

KI 98 DISK

Cells
Tissues
Organs

Embryonic Stem Cells as a Developmental Model in vitro

Guest Editors
A.M. Wobus, Gatersleben
K.R. Boheler, Baltimore, Md.

37 figures, 9 tables, 1999

KARGER
Basel · Freiburg · Paris · London · New York ·
New Delhi · Bangkok · Singapore · Tokyo · Sydney

Cover illustration from Cell Biology International 1996, Vol. 20, No. 8, pp. 579–587.

S. Karger
Medical and Scientific Publishers
Basel · Freiburg · Paris · London
New York · New Delhi · Bangkok
Singapore · Tokyo · Sydney

Drug Dosage
The authors and the publisher have exerted every effort to ensure that drug selection and dosage set forth in this text are in accord with current recommendations and practice at the time of publication. However, in view of ongoing research, changes in government regulations, and the constant flow of information relating to drug therapy and drug reactions, the reader is urged to check the package insert for each drug for any change in indications and dosage and for added warnings and precautions. This is particularly important when the recommended agent is a new and/or infrequently employed drug.

P.O. Box, CH–4009 Basel (Switzerland)
Printed in Switzerland on acid-free paper by
Reinhardt Druck, Basel
ISBN 3–8055–6975–0

KARGER

Fax+ 41 61 306 12 34
E-Mail karger@karger.ch
www.karger.com

Cells Tissues Organs

Contents Vol. 165, No. 3–4, 1999

KARGER

Fax+ 41 61 306 12 34
E-Mail karger@karger.ch
www.karger.com

Access to full text and tables of contents, including tentative ones for forthcoming issues: www.karger.com/journals/cto/cto_bk.htm

Cells
Tissues
Organs

Cells Tissues Organs 1999;165:129–130

Preface

A new era of developmental and molecular biology has grown from the successful derivation of undifferentiated embryonic stem (ES) cells from mouse embryos by Evans and Kaufmann [1981] and Martin [1981], and the application of gene targeting technology to ES cells by Thomas and Capecchi [1987]. As a direct consequence of these advances, hundreds of mutant mouse ('knock-out' and 'knock-in') strains have been established that pass genetic modifications to their offspring through germ-line transmission. ES cells not only give rise to all cell types in a mouse, but they differentiate in vitro into cells of endodermal, ectodermal and mesodermal lineages when cultured in three-dimensional aggregates called embryoid bodies. By using different methods to generate embryoid bodies, protocols have been established to differentiate ES cells into cardiogenic, myogenic, neurogenic, adipogenic, vascular and hematopoietic cells; consequently, more complex processes of cell differentiation, like vasculogenesis and angiogenesis, or neuromuscular interaction and synaptogenesis can be analyzed in culture. The use of ES cells for studying in vitro the regulation of cellular differentiation during early embryonic development by 'gain of function' and 'loss of function' analyses, the establishment of routine assays for pharmacological and toxicological screening procedures and, more recently, the selective differentiation of ES cells into specific somatic cell lineages are fields where ES cells have become more widely employed and accepted. Differentiation of ES cells in culture thus has developed into one of the most important cellular systems for the study of developmental biology.

The idea of a Special Issue to summarize our present knowledge of ES cell biology and their developmental potential in vitro arose at a Stem Cell Workshop held at an International Meeting of the European Tissue Culture Society in Mainz, October 1997. It seemed everyone was aware that it was only a matter of time until pluripotent stem cells of human origin would become available, and we discussed the potential prospects and problems of employing technologies with human ES cells at this meeting. In November 1998, the successful derivation of pluripotent embryo and primordial germ cell-derived cell lines originating from human embryonic and fetal tissue was reported by Thomson and Gearhart and their coworkers. The use of human ES cells for molecular and developmental biology and the exciting prospect of using these cells for human therapy seemed within grasp. It was only reasonable to assume that the techniques of in vitro differentiation which were established for mouse ES cells – together with the nuclear transfer technology described by Campbell and Wilmut [1997] – would be applied to pluripotent stem cells of human origin. Although some recent reviews devoted to specific fields like cardiogenesis are available [see Hescheler et al., 1997; Metzger et al., 1997; Wobus and Guan, 1998], a more comprehensive review of current uses and strategies with mouse ES cells was not. In this context, a Special Issue devoted to Embryonic Stem Cells as Developmental Model in vitro was envisioned.

Leading experts working with ES cells were contacted and invited to submit articles for this issue. Although not everyone could submit a manuscript within the designated time frame, this issue is a compilation of their and our efforts to present a comprehensive overview of the uses of ES cells in vitro. The first two chapters are dedicated to

KARGER
Fax + 41 61 306 12 34
E-Mail karger@karger.ch
www.karger.com

Accessible online at:
www.karger.com/journals/cto

molecular mechanisms regulating self-renewal and differentiation of ES cells (Burdon et al.) and the expression pattern of Oct-4, a key regulator of totipotency during the mammalian life cycle (Pesce et al.). The following articles review the present knowledge of ES cell-derived neurogenesis (Gottlieb and Huettner), adipogenesis (Dani) and hematopoiesis (Hole) on tissue-restricted gene and protein expression. To study developmental aspects of ion channel formation and signalling cascades during early developmental stages, one chapter is devoted to confocal imaging and patch clamp analyses of differentiated cells derived from ES cells (Hescheler et al.). The use of differentiation factors and signalling molecules for selective differentiation of ES cells will become more and more important; therefore, the effects of retinoic acid on stage-dependent differentiation of ES cells is summarized (Rohwedel et al.). Another article describes the use of ES cells in an in vitro embryotoxicity screening system, which is already part of a validation study of the European Union (Scholz et al.). Two chapters summarize the attempts to establish pluripotent cell lines from species other than mouse (Pain et al.), including the most recent strategies to introduce genetic modifications into livestock by nuclear transfer (Prelle et al.). The results of the nuclear transfer technology employed by Campbell and Wilmut [1997] led us to use the terms 'totipotent' for zygotes and blastomeres of early cleavage stage embryos, and 'pluripotent' for cells and cell lines derived from either the inner cell mass or primordial germ cells. As already mentioned, one of the most exciting areas of ES cell research is the isolation and exploitation of pluripotent stem cells of human origin and their possible use in tissue transplantations, a topic addressed in the final chapter of this issue (Boheler and Fiszman).

We are aware that the application of ES cell technology to pluripotent human ES cell lines has enormous ethical implications. International regulations for the use of stem cells and nuclear transfer techniques with human cells are needed to prevent abuse and overcome legal problems within and between countries. Perhaps the scientific information contained within this Special Issue will serve to inform and educate those interested parties involved in the debate on the uses and limitations of ES cells.

Finally, we are deeply grateful to the contributing authors for their participation and significant contributions to this issue. Although it was not possible to include all the interesting features of in vitro differentiated ES cells, it is hoped that this issue and the references included within will be valued by cell biologists working with pluripotent ES cells, both in vivo and in vitro, and those researchers who may wish to employ this highly versatile system in their studies in biotechnology and pharmacotoxicology, as well as in the establishment of techniques for tissue repair and cell therapy.

Anna M. Wobus, Gatersleben
Kenneth R. Boheler, Baltimore, Md.

References

Campbell, Wilmut (1997) Theriogenology *47*: 63–72.
Evans, Kaufman (1981) Nature *292*: 154–156.
Hescheler et al (1997) Cardiovasc Res *36*: 149–162.
Martin (1981) Proc Natl Acad Sci USA *78*: 7634–7638.
Metzger et al (1997) Trends Cardiovasc Med *7*: 63–68.
Thomas, Capecchi (1987) Cell *51*: 503–512.
Wobus, Guan (1998) Trends Cardiovasc Med *8*: 64–74.

Cells
Tissues
Organs

Cells Tissues Organs 1999;165:131–143

Accepted after revision: April 6, 1999

Signaling Mechanisms Regulating Self-Renewal and Differentiation of Pluripotent Embryonic Stem Cells

Tom Burdon Ian Chambers Craig Stracey Hitoshi Niwa Austin Smith

Centre for Genome Research, University of Edinburgh, UK

Key Words

Embryonic stem cells · Leukemia inhibitory factor · gp130 · STAT3 · Extracellular regulated kinase

Abstract

An ability to propagate pluripotent embryonic cells in culture is the foundation both for defined germline modification in experimental rodents and for future possibilities for broad-based cellular transplantation therapies in humans. Yet, the molecular basis of the self-renewing pluripotent phenotype remains ill-defined. The relationship between factors that influence embryonic stem cell propagation in vitro and mechanisms of stem cell regulation operative in the embryo is also uncertain. In this article we discuss the role of intracellular signalling pathways in the maintenance of pluripotency and induction of differentiation in embryonic stem cell cultures and the mammalian embryo.

Introduction

Mouse embryonic stem (ES) cells are continuous cell lines derived directly from the founder tissue of the embryo, the epiblast [Evans and Kaufman, 1981; Martin, 1981; Brook and Gardner, 1997]. They have apparently unlimited proliferative potential and can be propagated clonally with retention of non-transformed phenotype and euploid karyotype. Like the epiblast, ES cells are pluripotent, that is they are capable of multilineage differentiation. In vitro, ES cells can generate extra-embryonic cell types and derivatives of the three primary germ layers, ectoderm, mesoderm and endoderm. Most remarkably, in vivo, ES cells can be reincorporated into normal embryonic development and contribute to all developing cell lineages, including the germ line. ES cells thus represent capture in culture of the normally transient phase of embryonic pluripotency [Smith, 1992].

Abbreviations used in this paper

CNTF	ciliary neurotrophic factor
CT-1	cardiotrophin-1
ERK	extracellular regulated kinase
ES	embryonic stem cells
G-CSFR	granulocyte-colony-stimulating factor receptor
IL-6	interleukin-6
IRS	insulin receptor substrate
JAK	Janus kinase
LIF	leukemia inhibitory factor
MAPK	mitogen-activated protein kinases
OSM	oncostatin M
SH2	Src-homology 2
sIL-6R	soluble IL-6 receptor
STAT	signal transducer and activator of transcription

KARGER

Fax + 41 61 306 12 34
E-Mail karger@karger.ch
www.karger.com

1422–6405/99/1654–0131$17.50/0

Accessible online at:
www.karger.com/journals/cto

Drs. Tom Burdon or Austin Smith
Centre for Genome Research, University of Edinburgh
Kings Buildings, West Mains Road
Edinburgh, EH9 3JQ (UK)
Tel. +44 31 650 5890, Fax +44 31 667 0164, E-Mail tburdon@srv0.bio.ed.ac.uk

The ES cell genome can be manipulated with precision and relative ease [Smith et al., 1995], thereby enabling the generation of mice with defined genetic lesions. ES cell technology has consequently become a central research tool for determination of mammalian gene function and generation of animal models. To date, however, this technology has not been transferred to any species other than mouse despite considerable efforts [Gardner and Brook, 1997]. The barrier to establishing pluripotent ES cells from non-mouse embryos is not understood. Recent reports of primate and human embryo-derived cultures that share some of the properties of mouse ES cells are tantalising therefore [Thomson et al., 1998; Thomson and Marshall, 1998]. Human pluripotent cell lines would open up the possibility of cell therapy. In vitro stem cell differentiation could be used to produce tissue for cell replacement to treat a variety of genetic and acquired human ailments. If this prospect is to be realised, however, the human stem cells must exhibit the full proliferative potential and in vitro differentiation repertoire of mouse ES cells [Smith, 1998]. Furthermore, although progress to date has been dictated largely by empirical approaches, ultimately a rational appreciation of stem cell molecular biology is required to deliver clinical benefits with appropriate safeguards.

We describe here investigations into the signaling mechanisms by which cytokines sustain propagation of pluripotent ES cells. These studies have revealed both positive and negative effectors, suggesting that the decision between self-renewal and commitment is dictated by a balance of conflicting intracellular signals. Further understanding is needed, however, of how these opposing signals are generated and regulated and of the cross-talk between them. Such knowledge will inform and guide efforts to establish ES cell equivalents in other species and in addition may illuminate mechanisms of lineage commitment in the developing mammalian embryo.

Cytokine Dependence of Mouse ES Cells

Mouse ES cell lines are established by picking and expanding stem cell outgrowths derived from the inner cell masses of cultured blastocysts. Initial experimental protocols required that the stem cells were cultured on a monolayer of mouse embryo fibroblasts [Evans and Kaufman, 1981; Martin, 1981]. Without this feeder layer pluripotent cells could not be maintained, suggesting that the fibroblasts acted directly either to promote self-renewal or to suppress differentiation. Subsequent isolation of a soluble glycoprotein that prevents stem cell differentiation established that ES cell self-renewal is dependent on paracrine signal(s) [Smith and Hooper, 1987; Smith et al., 1988]. This factor is the cytokine leukaemia inhibitory factor (LIF) [Gearing et al., 1987; Williams et al., 1988].

LIF is a member of a family of related cytokines including interleukin-6 (IL-6), interleukin-11 (IL-11), oncostatin M (OSM), ciliary neurotrophic factor (CNTF) and cardiotrophin 1 (CT-1) [Taga and Kishimoto, 1997]. These factors have overlapping, pleiotropic effects on a variety of different cell types. In fact, OSM, CNTF and CT-1 can all suppress the differentiation of ES cells [Conover et al., 1993; Yoshida et al., 1994; Pennica et al., 1995]. In contrast, ES cells cannot be maintained by IL-6 or IL-11 since the specific receptors for these cytokines are not expressed on ES cells. However, if a soluble form of the IL-6 receptor is provided, IL-6 will support stem cell self-renewal [Nichols et al., 1994; Yoshida et al., 1994].

The basis for the shared activity of these cytokines is clear when the molecular structure of their receptor complexes is considered. LIF, OSM and CT-1 bind directly to a heterodimeric receptor complex containing two transmembrane glycoproteins, gp130 and LIFR [reviewed in Taga and Kishimoto, 1997; Heinrich et al., 1998] (fig. 1). CNTF also engages the heterodimer, but the interaction is mediated via binding to a GPi-anchored ligand binding protein, CNTFR. IL-6 acts independently of LIFR, through a hexameric receptor complex containing a homodimer of gp130 complexed with two ligand-bound IL-6 receptor modules. Thus, engagement of the gp130 receptor subunit is the common requirement for promotion of ES cell self-renewal by the LIF-related cytokines.

gp130 is a member of the cytokine receptor superfamily [Bazan, 1990]. In common with the other family members, the intracellular domain of gp130 has no intrinsic kinase activity. This function is provided via the association of Janus kinase (JAK) non-receptor tyrosine kinases with the membrane proximal, proline-rich, Box 1 region of gp130 [Darnell et al., 1994]. Ligand-induced oligomerisation of gp130 brings the associated JAKs into close proximity, causing their cross-phosphorylation and activation. JAKs 1, 2 and Tyk 2, are all reported to associate with gp130 [Lutticken et al., 1994; Stahl et al., 1994]. However, studies performed with mutant cell lines and mice lacking JAK 1 indicate that activity of this kinase in particular may be critical for eliciting downstream signaling events [Guschin et al., 1995; Rodig et al., 1998].

JAK activation is regarded as obligatory for tyrosine phosphorylation of the receptor and other downstream molecules. Receptor phosphotyrosines form binding sites for specific proteins containing phosphotyrosine binding motifs, such as the Src-homology 2 (SH2) domain. These bound molecules can then serve as further substrates for the JAK kinases. Whilst a degree of signaling specificity may reside in the JAK kinases, results of receptor domain swap experiments demonstrate that many key signals activated by cytokines are determined by the identity of the receptor tyrosine modules and the signal transducers that they recruit [Stahl et al., 1995].

A wide range of downstream effector molecules can be activated via gp130 [reviewed in Heinrich et al., 1998]. These include the signal transducer and activator of transcription (STAT) 1, 3 and 5 transcription factors, the tyrosine phosphatase SHP-2, the insulin receptor substrate (IRS) family proteins, IRS-1, IRS-2 and Gab1, the mitogen-activated protein kinases (MAPK), extracellular regulated kinases (ERK) 1 and 2, phosphoinositol-3 kinase, and the Src family tyrosine kinases, Hck, Btk and Fes. Some of these signals may be cell type-specific and their contribution to eliciting biological effects downstream of gp130 in many cases remains to be determined. It has been clearly established, however, that activation of the STAT and ERK signaling pathways do play essential roles in mediating the biological responses to cytokines acting through gp130. Activation of STAT3 is required for gp130-dependent differentiation of the M1 myeloid cells [Minami et al., 1996; Nakajima et al., 1996] and promotes the differentiation of astrocytes [Bonni et al., 1997]. In the BAF3 pro-B-cell line, STAT3 induces antiapoptotic factors and suppresses antiproliferative signals [Fukada et al., 1996, 1998]. On the other hand, stimulation of the MAPK pathway via gp130 has been shown to promote neurite outgrowth [Ihara et al., 1997], cardiomyocyte survival [Sheng et al., 1997], and BAF3 cell proliferation [Fukada et al., 1996].

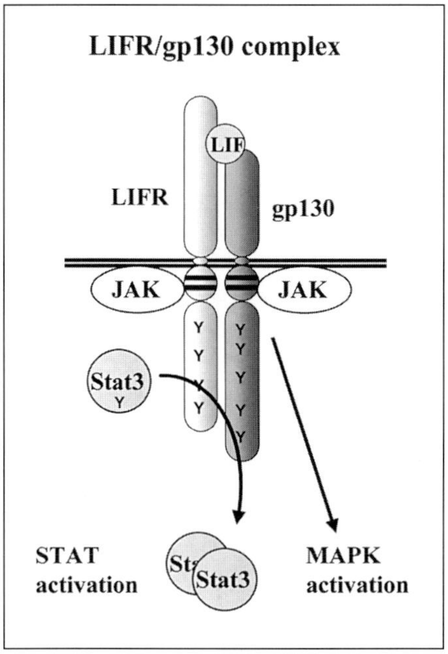

Fig. 1. The LIFR/gp130 signaling complex. LIF binds to LIFR and gp130, thereby bringing together the receptor-associated JAK kinases. Subsequent JAK activation and phosphorylation of receptor tyrosines then lead to stimulation of both the STAT and MAPK signaling pathways.

Activation of STAT3

STAT3 was first identified as a component of the IL-6-induced DNA binding activity, acute phase response factor [Akira et al., 1994; Zhong et al., 1994]. Subsequently, it was shown to be activated by a variety of cytokine receptors, receptor tyrosine kinases and non-receptor tyrosine kinases. Seven mammalian STAT genes have been identified and encode proteins with a similar overall structure [Darnell, 1997]. These proteins contain a highly conserved amino terminus that mediates interactions between activated STAT dimers, a centrally located DNA binding region, an SH2 domain, an adjacent, positionally conserved regulatory tyrosine and a carboxy terminal region containing a transactivation domain (fig. 2). The most striking feature of STATs, the SH2 domain, is not found in any other known family of transcription factors. This enables STATs to link activation of receptors at the cell surface directly with transcriptional events in the nucleus.

The SH2 domain mediates the selective recruitment of STATs to activated receptor complexes. STAT3, for instance, has been shown to bind receptors carrying YpXXQ motif (where X represents any amino acid) [Stahl et al., 1995; Gerhartz et al., 1996]. Receptor-associated STATs are then subject to the attentions of tyrosine kinases, becoming phosphorylated on the conserved tyrosine, immediately adjacent to the SH2 domain [Wen et al., 1995]. Upon dissociation from the receptor, STATs dimerise through reciprocal interactions involving the SH2 domains and phosphotyrosines. Activation of STAT3 is often accompanied by phosphorylation of STAT1, with which it can heterodimerise. The balance of homodimers and heterodimers is receptor- and cell type-dependent and can be influenced by the level of activating

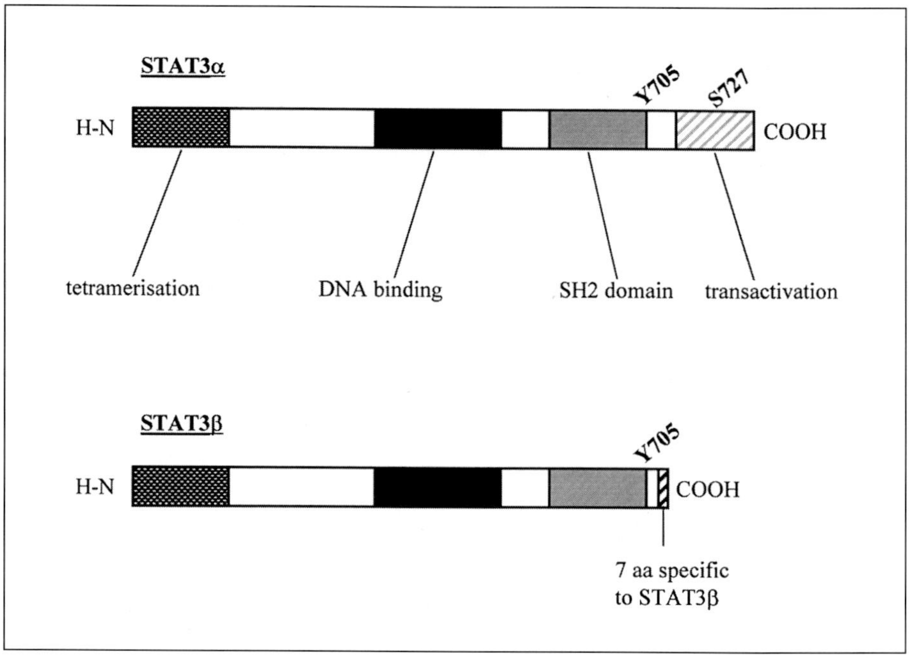

Fig. 2. Functional domains of STAT3.

ligand [Darnell, 1997]. STAT dimers then rapidly translocate into the nucleus, where they bind to DNA and regulate transcription of target genes. All STAT transcription factors (with the exception of STAT2) can bind to variants of the palindromic consensus sequence TTCnnnGAA, presumably reflecting the symmetry of the STAT dimer [Hoey and Schindler, 1998].

STAT3β is a naturally occurring splice variant of the full-length STAT3 (STAT3α) in which the carboxy terminal 55 residues encompassing the transactivation domain are replaced with an unrelated 7-residue peptide [Schaefer et al., 1995] (see fig. 2). The truncated STAT3β protein retains tyrosine 705 and can therefore be activated, forming dimers capable of binding to the STAT DNA recognition site [Schaefer et al., 1997]. Although STAT3β can function as an inhibitor of full-length STAT3, under some circumstances it transactivates STAT3 reporter genes, possibly directly or through an association with the Jun B transcription factor [Schaefer et al., 1995; Sasse et al., 1997; Schaefer et al., 1997]. Interestingly, following cytokine stimulation, the tyrosine phosphorylated form of STAT3β persists for longer than activated STAT3α, indicating that turnover of activated STAT3 is regulated through its carboxy terminus.

The transactivation domain of STAT3α is also subject to ligand-dependent phosphorylation of residue serine 727 [Wen et al., 1995]. A mutant STAT3, lacking S727 exhibits reduced transcriptional activation of STAT-dependent reporter genes and inhibits STAT3-dependent transformation [Wen et al., 1995; Bromberg et al., 1998]. The phosphorylation of S727 may therefore contribute to the transactivation function of STAT3. Interestingly, S727 is situated within a recognised consensus recognition site for MAPK [Chung et al., 1997], indicating that phosphorylation of S727 may represent a direct mechanism for integrating MAPK and STAT signaling pathways. The kinase responsible for serine phosphorylation in vivo has not yet been identified, however.

STAT3 and Regulation of ES Cell Self-Renewal

STAT3 is the principal STAT factor activated in ES cells stimulated with either IL-6 plus soluble IL-6 receptor (sIL-6R) (fig. 3), or with LIF [Hocke et al., 1995]. The majority of STAT DNA binding complexes are comprised of STAT3 homodimers (fig. 3), although low levels of activated STAT1 can also be detected [Hocke et al., 1995; Starr et al., 1997]. On the basis of functional assays in somatic cell lines and in vitro peptide binding experiments, four tyrosines within the cytoplasmic domain of gp130 conforming to the YXXQ STAT3 consensus binding motif represent potential STAT docking sites [Stahl et al., 1995; Gerhartz et al., 1996]. The most membrane proximal site is contained within the conserved box 3 domain (Y126, situated 126 amino acids from the cyto-

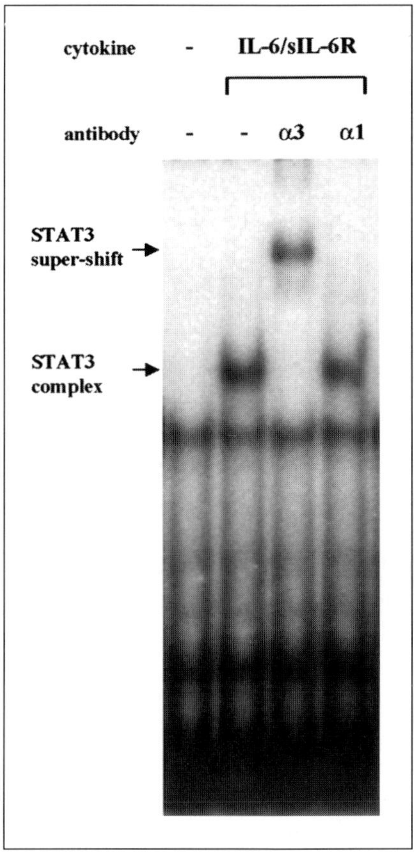

Fig. 3. gp130-dependent activation of STAT3 in ES cells. Nuclear extracts prepared from uninduced ES cells or cells treated with IL-6 plus sIL-6R were incubated with a ^{32}P-labelled oligonucleotide containing a STAT DNA recognition site and fractionated on a non-denaturing polyacrylamide gel. The induced STAT DNA binding complex is composed mainly of STAT3 (homodimers), since it is completely supershifted by antibodies specific for STAT3 (α3), but unaffected by antibodies recognising STAT1 (α1).

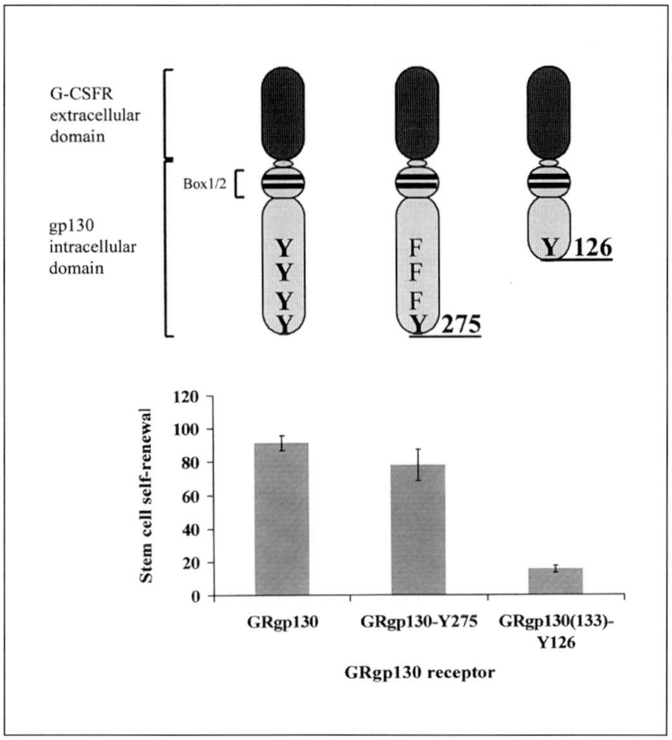

Fig. 4. Amino-terminal STAT3 binding site in gp130 is sufficient for directing ES cell self-renewal. ES cell transfectants expressing G-CSFR/gp130 receptor chimaeras were plated at medium density (5,000 cells per well of a 24-well dish) and cultured for 6 days in medium containing either no cytokine, G-CSF or IL-6 plus sIL-6R. Stem cell self-renewal was measured by assaying stem cell-specific expression of a chromosomal *Oct4-lacZ* reporter gene. The results presented in the graph represent the level of β-galactosidase activity induced by G-CSF, expressed as a percentage of that obtained with IL-6 plus sIL-6R. The data shown is the mean values (±SEM) obtained from three independent transfectants for each chimaeric receptor construct.

plasmic face of the membrane). The next motif is located at 173 amino acids (Y173) and the remaining two are situated adjacent to each other at the very carboxy terminus of the receptor (Y265, Y275).

In order to assess the contribution of STAT3 to ES cell self-renewal, chimaeric cytokine receptors consisting of the extracellular domain of the granulocyte-colony-stimulating factor receptor (G-CSFR) fused to either the wild type or mutated forms of the intracellular domain of gp130 were stably introduced into ES cells. The recipient cell line used in these studies contains a β-galactosidase gene integrated into the genomic locus encoding the essential stem cell-specific transcription factor Oct4 [Nichols et al., 1998]. As a result, expression of the *lacZ* reporter gene is restricted to undifferentiated ES cells and provides a reliable measure of stem cell self-renewal [Dani et al., 1998].

G-CSFR is a cytokine receptor that forms homodimers in response to administration of G-CSF. ES cells do not express the endogenous G-CSFR and no response to G-CSF is evident either biochemically or at the level of self-renewal. This system, therefore, allows the functional analysis of gp130-dependent signals independently from the contribution of endogenous cytokine receptors. In fact, the chimaeric receptor containing the complete cytoplasmic domain of the gp130 receptor is fully capable of directing ES self-renewal [Starr et al., 1997; Niwa et al., 1998], demonstrating that the hybrid molecule reproduces the essential functions of endogenous gp130 in ES cells (fig. 4).

Substitution of all four STAT3 binding tyrosines with phenylalanine within this chimaeric receptor abolishes both activation of STAT3 and the capacity of the receptor to prevent the differentiation of ES cells [Niwa et al., 1998]. Nevertheless, the receptor still engages productively with G-CSF, as is evident by its ability to activate the MAPK pathway. Interestingly, the four YXXQ binding sites on gp130 are not equally effective at activating STAT3. Loss of the two proximal sites, Y126, and Y173, does not appreciably diminish the activation of STAT3 or self-renewal. In contrast, substitution of both Y265 and Y275 with phenylalanine significantly reduces activation of STAT3 and impairs the ability of the receptor to promote ES cell self-renewal. This suggests that in ES cells only the terminal tyrosine motifs of gp130 are fully effective in activating STAT3. This conclusion is supported by the findings that a chimaeric receptor in which only the terminal Y275 STAT3 binding site is intact is capable of blocking the differentiation of ES cells (fig. 4). In comparison, a truncated chimaeric receptor containing only the most proximal YXXQ motif, which is reported to be effective biochemically and biologically in M1 or BAF3 cells [Fukada et al., 1996; Yamanaka et al., 1996], is incapable of delivering a full self-renewal response in ES cells.

The molecular basis for this disparity is not understood. Such differences could be trivial and artifactual: it may be an inherent feature of continual passaging that cells are selected for transformation or increased sensitivity to extracellular signals. Such cells could be particularly adept at rescuing otherwise weak STAT3 binding sites from signaling obscurity. This explanation, however, provides no insight into the presence of four potential STAT3 binding sites in gp130. All four sites are conserved between mouse and man suggesting that there have been selective pressures for retaining their functions and that they play important individual or additive roles under specific circumstances or in particular cell types. Interestingly, the gp130-related cytokine receptor, G-CSFR, expression of which is limited to myeloid cell types, contains only one consensus STAT3 binding site. Multimerisation of the STAT3 docking sites on gp130 may have been instrumental in its recruitment to functions in a wide variety of different cell types, including epiblast cells. One possibility is that differential association of other signaling complexes, for example through recruitment to Y118 (see below), may influence the accessibility of particular YXXQ motifs.

It is also conceivable that the two terminal tyrosines in gp130 have functions in ES cells other than their ability to engage STAT3 and these additional signals are necessary for self-renewal. STAT1 can be activated via gp130 and the terminal two tyrosines have been reported to be particularly important in this process [Gerhartz et al., 1996]. It is unlikely, however, that STAT1 is required for self-renewal since STAT1$^{-/-}$ ES cells can be propagated in a LIF-dependent manner [Durbin et al., 1996].

Inhibition of STAT3 signaling directly, through the use of interfering mutant forms of the transcription factor, provides the best evidence for an essential role of STAT3 in ES cell self-renewal. Expression of STAT3F, a mutant in which the tyrosine 705 required for dimerisation is substituted with phenylalanine, blocks STAT3 activity presumably by either competing for receptor binding sites, or by forming non-functional complexes with endogenous STAT3 [Kaptein et al., 1996; Minami et al., 1996]. Significantly, ES cell transfectants that constitutively express high levels of STAT3F from a stably integrated transgene cannot be propagated [Boeuf et al., 1997; Niwa et al., 1998]. The basis for this observation became apparent by using a polyoma-based episomal expression system, in which transfection efficiencies 100-fold higher than conventional protocols can routinely be attained [Gassmann et al., 1995; Niwa et al., 1998]. Inspection of the resulting ES cell transfectants established that high levels of STAT3F expression cause ES cells to differentiate [Niwa et al., 1998]. Importantly, whereas co-expression of STAT3 with STAT3F restores stem cell self-renewal, overexpression of STAT1 does not, indicating that this closely related STAT family member cannot functionally substitute for STAT3. Furthermore, conditional expression of an integrated, tetracycline-inducible STAT3F transgene also blocks ES cell self-renewal [Niwa et al., 1998]. In conjunction with the results obtained with the receptor chimaeras, these data establish that activation of STAT3 is critical for gp130-dependent self-renewal of ES cells.

Whereas the evidence for a role of STAT3 in ES cells is compelling, the importance of gp130 signaling in promoting stem cell self-renewal in the early embryo is unclear. Although LIF, LIFR and more significantly, gp130, are all expressed appropriately within the blastocyst [Nichols et al., 1996], deletion of any of these individual components through homologous recombination does not prevent formation of the epiblast, nor appreciably affect egg cylinder morphogenesis or gastrulation [Stewart et al., 1992; Li et al., 1995; Yoshida et al., 1996]. STAT3$^{-/-}$ embryos, in contrast, are reported to die early in development, at about embryonic day 7.5 [Takeda et al., 1997]. At this stage the embryos have formed an egg cylinder but are reported to be unable to progress through gastrulation. In situ hybridisation experiments indicate that STAT3 transcription is first expressed in the visceral endoderm [Takeda et al., 1997]. On this basis, it was proposed that the early developmental failure is caused by a defect in the formation of extraembryonic tissues [Takeda et al., 1997]. However, when STAT3$^{-/-}$ blastocysts are cultured in vitro the inner cell mass does not thrive, consistent with a requirement for STAT3 in the maintenance of ES cells. The discrepancy between the phenotype of the gp130 and STAT3 null embryos and the importance of gp130 signaling in vitro, however, does necessitate careful reconsideration of the relationship between epiblast stem cells in vivo and ES cells. If STAT3 is dispensable for expansion of the epiblast it is likely that other signals operate in vivo. The report of a gp130-independent factor secreted by extraembryonic cells, that can suppress differentiation of ES cells but does not activate STAT3, supports the existence of alternative self-renewal pathways [Dani et al., 1998].

Although STAT3 is not required for the development of the epiblast, it could function in an adaptive mechanism for maintaining stem cell pluripotency when early embryonic development is perturbed. Such a situation occurs naturally in rodents when a lactating female conceives while still nursing. Under these circumstances there is a delay in implantation of the blastocysts, termed diapause. This can also be induced experimentally by ovariectomising a pregnant mouse shortly after conception [Hogan et al., 1994]. In this condition, implantation of blastocysts in the uterus wall is blocked. The blastocysts are viable, however, remaining in a state of developmental arrest within the uterus for up to 4 weeks. When transplanted into the uterus of a surrogate mouse, or following administration of oestrogen, they can implant and reinitiate normal development. Aside from a slight increase in the number of cells in the epiblast, little is known about maintenance of stem cells during delayed implantation. Interestingly, however, implantation delay appears to enhance the frequency of ES cell derivation [Evans and Kaufman, 1981; Gardner and Brook, 1997]. STAT3 may therefore contribute to maintaining the pluripotency of stem cells during diapause, which would provide a rationale for its role in promoting ES cell self-renewal in vitro.

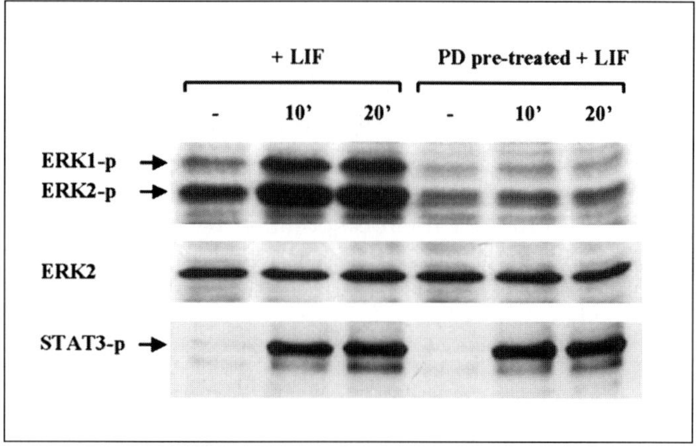

Fig. 5. LIF-dependent ERK activation in ES cells. ES cells either pretreated overnight with the MEK inhibitor PD098059 (PD, 50 μM), or untreated, were stimulated with LIF for 0, 10 and 20 min and analysed by Western blotting. Activation of ERKs 1 and 2 (recognised by a phospho-ERK-specific antibody) is rapidly induced after treatment with LIF. Significantly, the pretreatment of ES cells with PD098059 blocks ERK activation but does not affect the level of ERK protein or impair STAT3 activation (recognised by a phospho-STAT3-specific antibody).

gp130-Dependent Activation of SHP-2 and ERK Signaling

The other major class of signals shown to be activated downstream of gp130 are MAP kinases. ERKs 1 and 2 are activated in ES cells stimulated with LIF or IL-6 plus sIL-6R [Ernst et al., 1996; Boeuf et al., 1997] (fig. 5). These signal transducers attract interest because of their well-documented functions in regulating the proliferation and survival of somatic cell types [Lloyd, 1998]. A possible role for this pathway in ES self- renewal has been implied by observations that expression of upstream activators of ERKs, either constitutively activated forms of Hck, a Src family kinase that may be upstream of Ras, or a hyperactive form of Ras itself, reduced the requirement of ES cells for LIF [Ernst et al., 1994]. This issue is contentious, however, since over-expression of, or constitutive activation of, proto-oncogenic tyrosine kinases can have unforeseen regulatory consequences. There is evidence that Src itself can also promote ES cell self-renewal [Boulter et al., 1991], and this could be explained through its established ability to activate STAT3 [Yu et al., 1995; Cao et al., 1996]. Furthermore, expression of a constitutively active form of Ras in ES cells is reported to direct visceral endoderm differentiation [Cheng et al., 1998].

One established route for receptor-mediated activation of ERKs in somatic cells is through the cytosolic protein tyrosine phosphatase SHP-2. This ubiquitously expressed phosphatase is homologous to the *Drosophila* protein corkscrew and closely related to the mammalian haematopoietic cell phosphatase SHP-1 [reviewed in Neel and Tonks, 1997; Van Vactor et al., 1998]. The common structural features of this family of proteins are two tandemly arranged SH2 domains in the amino terminus and a catalytic domain within the carboxy terminus. The SH2 domains recruit SHP-2 to activated receptors and potential substrates. In addition, engagement of the amino-terminal SH2 motif, by preventing its intramolecular association with the phosphatase domain, increases the catalytic activity of SHP-2 [Hof et al., 1998].

SHP-2 is recruited to gp130 through a single phosphorylated tyrosine, located 118 amino acids from the membrane within the box 3 region [Stahl et al., 1995]. Mutation of Y118 both prevents binding of SHP-2 and blocks the activation of ERKs via gp130 [Fukada et al., 1996]. Two alternative, but not necessarily exclusive, mechanisms may explain how SHP-2 couples to the ERK pathway. Upon recruitment, SHP-2 becomes phosphorylated on two tyrosines situated close to carboxy terminus [Bennett et al., 1994; Fukada et al., 1996]. One of these, Y542, constitutes a binding site for the adaptor protein Grb2. This interaction provides a relatively direct mechanism by which SHP-2 can be coupled through Grb2 and Sos to Ras, thereby activating the ERK signaling pathway [Li et al., 1994]. Alternatively, activation of ERKs can occur through the association of SHP-2 with a potential substrate, Gab1, an IRS-like multifunctional adaptor protein [Holgado-Madruga et al., 1996; Takahashi-Tezuka et al., 1998]. Genetic and biochemical evidence in *Drosophila* identifies a Gab1-related protein, Dos, as a necessary substrate for signaling through the corkscrew protein [Herbst et al., 1996; Raabe et al., 1996]. Furthermore, experiments performed with mammalian cells suggest that Gab1 and SHP-2 are recruited to gp130, forming a complex with and activating phosphoinositol-3-kinase, which in turn stimulates ERK signaling via a Ras-dependent pathway [Takahashi-Tezuka et al., 1998].

In addition to its documented role as positive effector, recent studies suggests that SHP-2 can also have a negative role in signaling. Indeed, this was hinted at by its homology to SHP-1, an established negative regulator, and its association with CTL-4, an inhibitor of T cell receptor activity [Marengere et al., 1996]. Direct evidence for a negative role has now been provided by studies of gp130 signaling. Elimination of the SHP-2 binding site in gp130 or overexpression of catalytically inactive SHP-2 mutants results in enhanced activation of JAK kinases, prolonged phosphorylation of gp130 and STAT3, and increased transcription of STAT3 target genes [Symes et al., 1997; Kim et al., 1998; Servidei et al., 1998]. These results have been taken to indicate that SHP-2 may directly dephosphorylate JAKs or the receptor. However, the reported antagonism of STAT3 activity by ERK signaling suggests that enhanced STAT3 activity associated with the absence of SHP-2 function could also result from the loss of signals downstream from the phosphatase [Bonni et al., 1997; Ihara et al., 1997].

Role of SHP-2/ERK Signaling in ES Cell Self-Renewal

SHP-2, a Gab1-related protein and ERKs 1 and 2 are activated in ES cells stimulated with IL-6 plus sIL-6R (fig. 5) [Burdon et al., 1999]. To examine the role of this pathway, a chimaeric G-CSFR/gp130 receptor in which the Y118 SHP-2 binding site is mutated to phenylalanine was stably introduced into ES cells. This receptor does not engage SHP-2, yet is fully capable of directing the propagation of ES cells [Burdon et al., 1999]. Indeed, the mutant receptor exhibits increased signaling activity as evident by prolonged activation of STAT3 and enhanced ES cell self-renewal.

This lack of a requirement for SHP-2 activity is supported by results obtained from ES cells in which both copies of the phosphatase gene have been mutated by homologous recombination [Saxton et al., 1997; Qu and Feng, 1998]. The targeted deletion eliminates 65 amino acids from the amino terminal SH2 domain and increases the constitutive activity of the phosphatase [Qu et al., 1997]. However, the mutant SHP-2 protein (SHP-$2^{\Delta 46-110}$) is expressed at reduced levels and without a functional amino terminal SH2 domain would not be expected to interact effectively with many of its target proteins. This contention is supported by the reported deregulation of signaling via several different types of receptor in *SHP-2*$^{\Delta 46-110}$ cells [Saxton et al., 1997]. Nevertheless, the mutant ES cells are viable and in fact exhibit increased sensitivity to LIF [Qu and Feng, 1998], consistent with the notion that SHP-2 acts primarily as negative regulator of self-renewal signaling in ES cells [Burdon et al., 1999].

The absence of an essential requirement for SHP-2 function in ES cells implies that ERK signaling does not contribute directly to stem cell self-renewal. Indeed, minimal activation of ERKs is detected following stimulation through the mutant G-CSFR/gp130 Y118F receptor [Burdon et al., 1999]. Furthermore treatment of ES cells with blocking concentrations of PD098059, an inhibitor of the MEK enzymes that activate ERKs (fig. 5), does not impair the propagation of ES cells. In fact, ES cells cultured in PD098059 remain pluripotent, contributing to

1996; Bonni et al., 1997]. Given the evidence that ERK activation is likely to play a role in the establishment of differentiated cell lineages, the role of STAT3 in ES cells may include blocking the actions of this signal (fig. 6). This could be effected at the level of activation, or STAT3 and ERK signaling pathways may converge and compete at the level of gene expression. STAT3 may act by repressing transcription of differentiation genes, whereas ERK-dependent factors may induce their expression.

Further investigations into how STAT3 and ERK signals regulate ES cell self-renewal are likely to provide valuable insights into understanding the regulation of stem cell pluripotency. On the basis of current evidence it appears that the balance between these two pathways may determine the choice between self-renewal and differentiation (fig. 6). Importantly, therefore, selective inhibition or activation of these signals and their targets may allow greater control over the propagation, manipulation and differentiation of ES cells.

Acknowledgements

The work reported from the authors' laboratory was supported by the Biotechnology and Biological Sciences Research Council of the United Kingdom and the Human Frontiers Science Program Organisation.

References

Akira, S., Y. Nishio, M. Inoue, X.J. Wang, S. Wei, T. Matsusaka, K. Yoshida, T. Sudo, M. Naruto, T. Kishimoto (1994) Molecular cloning of APRF, a novel IFN-stimulated gene factor 3 p91-related transcription factor involved in the gp130-mediated signaling pathway. Cell *77:* 63–71.

Arrandale, J.M., A. Gore-Willse, S. Rocks, J.M. Ren, J. Zhu, A. Davis, J.N. Livingston, D.U. Rabin (1996) Insulin signaling in mice expressing reduced levels of Syp. J Biol Chem *271:* 21353–21358.

Arsenian, S., B. Weinhold, M.U.R. Oelgeschlager, A. Nordheim (1998) Serum response factor is essential for mesoderm formation during mouse embryogenesis. EMBO J *17:* 6289–6299.

Bazan, J.F. (1990) Structural design and molecular evolution of a cytokine receptor superfamily. Proc Natl Acad Sci USA *87:* 6934–6938.

Bennett, A.M., T.L. Tang, S. Sugimoto, C.T. Walsh, B.G. Neel (1994) Protein-tyrosine-phosphatase SHPTP2 couples platelet-derived growth-factor receptor-beta to Ras. Proc Natl Acad Sci USA *91:* 7335–7339.

Boeuf, H., H. Hauss, F. De Graeve, N. Baran, C. Kedinger (1997) Leukemia inhibitory factor-dependent transcriptional activation in embryonic stem cells. J Cell Biol *138:* 1207–1217.

Bonni, A., Y. Sun, M. Nadal-Vicens, A. Bhatt, D.A. Frank, I. Rozovsky, N. Stahl, G.D. Yancopoulos, M.E. Greenberg (1997) Regulation of gliogenesis in the central nervous system by the JAK-STAT signaling pathway. Science *278:* 477–483.

Boulter, C.A., A. Aguzzi, R.L. Williams, E.F. Wagner, M.J. Evans, R. Beddington (1991) Expression of v-src induces aberrant development and twinning in chimaeric mice. Development *111:* 357–366.

Bromberg, J.F., C.M. Horvath, D. Besser, W.W. Lathem, J.E. Darnell, Jr. (1998) Stat3 activation is required for cellular transformation by v-src. Mol Cell Biol *18:* 2553–2558.

Brook, F.A., R.L. Gardner (1997) The origin and efficient derivation of embryonic stem cells in the mouse. Proc Natl Acad Sci USA *94:* 5709–5712.

Burdon, T., C. Stracey, I. Chambers, J. Nichols, A. Smith (1999) Suppression of SHP-2 and ERK signaling promotes self-renewal of mouse embryonic stem cells. Dev Biol *210:* 30–43.

Cao, X., A. Tay, G.R. Guy, Y.H. Tan (1996) Activation and association of stat3 with src in v-src-transformed cell lines. Mol Cell Biol *16:* 1595–1603.

Cheng, A.M., T.M. Saxton, R. Sakai, S. Kulkarni, G. Mbamalu, W. Vogel, C.G. Tortorice, R.D. Cardiff, J.C. Cross, W.J. Muller, T. Pawson (1998) Mammalian Grb2 regulates multiple steps in embryonic development and malignant transformation. Cell *95:* 793–803.

Chung, J., E. Uchida, T.C. Grammer, J. Blenis (1997) STAT3 serine phosphorylation by ERK-dependent and -independent pathways negatively modulates its tyrosine phosphorylation. Mol Cell Biol *17:* 6508–6516.

Conover, J.C., N.Y. Ip, W.T. Poueymirou, B. Bates, M.P. Goldfarb, T.M. DeChiara, G.D. Yancopoulos (1993) Ciliary neurotrophic factor maintains the pluripotentiality of embryonic stem cells. Development *119:* 559–565.

Dani, C., I. Chambers, S. Johnstone, M. Robertson, B. Ebrahimi, M. Saito, T. Taga, M. Li, T. Burdon, J. Nichols, A. Smith (1998) Paracrine induction of stem cell renewal by LIF-deficient cells: A new ES cell regulatory pathway. Dev Biol *203:* 149–162.

Darnell, J.E., Jr. (1997) STATs and gene regulation. Science *277:* 1630–1635.

Darnell, J.E., Jr., I.M. Kerr, G.R. Stark (1994) Jak-STAT pathways and transcriptional activation in response to IFNs and other extracellular signaling proteins. Science *264:* 1415–1421.

Durbin, J.E., R. Hackenmiller, M.C. Simon, D.E. Levy (1996) Targeted disruption of the mouse Stat1 gene results in compromised innate immunity to viral disease. Cell *84:* 443–450.

Ernst, M., D.P. Gearing, A.R. Dunn (1994) Functional and biochemical association of Hck with the LIF/IL-6 receptor signal transducing subunit gp130 in embryonic stem cells. EMBO J *13:* 1574–1584.

Ernst, M., Oates, A., Dunn, A. R. (1996). Gp130-mediated signal transduction in embryonic stem cells involves activation of Jak and Ras/mitogen activated protein kinase pathways. J Biol Chem *271:* 30136–30143.

Evans, M.J., M. Kaufman (1981) Establishment in culture of pluripotential cells from mouse embryos. Nature *292:* 154–156.

Fukada, T., M. Hibi, Y. Yamanaka, M. Takahashi-Tezuka, Y. Fujitani, T. Yamaguchi, K. Nakajima, T. Hirano (1996) Two signals are necessary for cell proliferation induced by a cytokine receptor gp130: Involvement of STAT3 in anti-apoptosis. Immunity *5:* 449–460.

Fukada, T., T. Ohtani, Y. Yoshida, T. Shirogane, K. Nishida, K. Nakajima, M. Hibi, T. Hirano (1998) STAT3 orchestrates contradictory signals in cytokine-induced G1 to S cell-cycle transition. EMBO J *17:* 6670–6677.

Gardner, R.L., F.A. Brook (1997) Reflections on the biology of embryonic stem (ES) cells. Int. J Dev Biol *41:* 235–243.

Gassmann, M., G. Donoho, P. Berg, P. Maintenance of an extrachromosomal plasmid vector in mouse embryonic stem cells. Proc Natl Acad Sci USA *92:* 1292–1296.

Gearing, P.D., N.M. Gough, J.A. King, D.J. Hilton, N.A. Nicola, R.J. Simpson, E.C. Nice, A. Kelso, D. Metcalf (1987) Molecular cloning and expression of cDNA encoding a murine myeloid leukemia inhibitory factor (LIF). EMBO J *6:* 3995–4002.

Gerhartz, C., B. Heesel, J. Sasse, U. Hemmann, C. Landgraf, J. Schneider-Mergener, F. Horn, P.C. Heinrich, L. Graeve (1996) Differential activation of acute phase response factor/STAT3 and STAT1 via the cytoplasmic domain of the interleukin 6 signal transducer gp130. I. Definition of a novel phosphotyrosine motif mediating STAT1 activation. J Biol Chem *271:* 12991–12998.

Guschin, D., N. Rogers, J. Briscoe, B. Witthuhn, D. Watling, F. Horn, S. Pellegrini, K. Yasukawa, P. Heinrich, G.R. Stark, et al (1995) A major role for the protein tyrosine kinase JAK1 in the JAK/STAT signal transduction pathway in response to interleukin-6. EMBO J *14:* 1421–1429.

Heinrich, P.C., I. Behrmann, G. Muller-Newen, F. Schaper, L. Graeve (1998) Interleukin-6-type cytokine signalling through the gp130/Jak/STAT pathway. Biochem J *334:* 297–314.

Herbst, R., R.M. Carroll, J.D. Allard, J. Schilling, T. Raabe, M.A. Simon (1996) Daughter of sevenless is a substrate of the phosphotyrosine phosphatase Corkscrew and functions during sevenless signaling. Cell *85:* 899–909.

Hirano, T., K. Nakajima, M. Hibi (1997) Signaling mechanisms through gp130: A model of the cytokine system. Cytokine Growth Factor Rev *8:* 241–252.

Hocke, G.M., M.Z. Cui, G.H. Fey (1995) The LIF response element of the α2 macroglobulin gene confers LIF-induced transcriptional activation in embryonal stem cells. Cytokine *7:* 491–502.

Hoey, T., U. Schindler (1998) STAT structure and function in signaling. Curr Opin Genet Dev *8:* 582–587.

Hof, P., S. Pluskey, S. Dhe-Paganon, M.J. Eck, S.E. Shoelson (1998) Crystal structure of the tyrosine phosphatase SHP-2. Cell *92:* 441–450.

Hogan, B., R. Beddington, F. Constantini, E. Lacy (1994) Manipulating the mouse embryo: A Laboratory Manual, ed 2. Cold Spring Harbor, Cold Spring Harbor Laboratory Press.

Holgado-Madruga, M., D.R. Emlet, D.K. Moscatello, A.K. Godwin, A.J. Wong (1996) A Grb2-associated docking protein in EGF- and insulin-receptor signalling. Nature *379:* 560–564.

Ihara, S., K. Nakajima, T. Fukada, M. Hibi, S. Nagata, T. Hirano, Y. Fukui (1997) Dual control of neurite outgrowth by STAT3 and MAP kinase in PC12 cells stimulated with interleukin-6. EMBO J *16:* 5345–5352.

Kaptein, A., V. Paillard, M. Saunders (1996) Dominant negative stat3 mutant inhibits interleukin-6-induced Jak-STAT signal transduction. J Biol Chem *271:* 5961–5964.

Kim, H., T.S. Hawley, R.G. Hawley, H. Baumann (1998) Protein tyrosine phosphatase 2 (SHP-2) moderates signaling by gp130 but is not required for the induction of acute-phase plasma protein genes in hepatic cells. Mol Cell Biol *18:* 1525–1533.

LaBonne, C., B. Burke, M. Whitman (1995) Role of MAP kinase in mesoderm induction and axial patterning during Xenopus development. Development *121:* 1475–1486.

Li, M., M. Sendtner, A. Smith (1995) Essential function of LIF receptor in motor neurons. Nature *378:* 724–727.

Li, W., R. Nishimura, A. Kashishian, A.G. Batzer, W.J. Kim, J.A. Cooper, J. Schlessinger (1994) A new function for a phosphotyrosine phosphatase: Linking GRB2-Sos to a receptor tyrosine kinase. Mol Cell Biol *14:* 509–517.

Lloyd, A.C. (1998) Ras versus cyclin-dependent kinase inhibitors. Curr Opin Genet Dev *8:* 43–48.

Lutticken, C., U.M. Wegenka, J. Yuan, J. Buschmann, C. Schindler, A. Ziemiecki, A.G. Harpur, A.F. Wilks, K. Yasukawa, T. Taga, T. Kishimoto, G. Barbieri, S. Pellegrini, M. Sendtner, P.C. Heinrich, F. Horn (1994) Association of transcription factor APRF and protein kinase Jak1 with the interleukin-6 signal transducer gp130. Science *263:* 89–92.

Mac Auley, A., Z. Werb, P.E. Mirkes (1993) Characterization of the unusually rapid cell cycles during rat gastrulation. Development *117:* 873–883.

Marengere, L.E., P. Waterhouse, G.S. Duncan, H.W. Mittrucker, G.S. Feng, T.W. Mak (1996) Regulation of T cell receptor signaling by tyrosine phosphatase SYP association with CTLA-4. Science *272:* 1170–1173.

Martin, G.R. (1981) Isolation of a pluripotent cell line from early mouse embryos cultured in medium conditioned by teratocarcinoma stem cells. Proc Natl Acad Sci USA *78:* 7634–7638.

Minami, M., M. Inoue, S. Wei, K. Takeda, M. Matsumoto, T. Kishimoto, A. Akira (1996) STAT3 activation is a critical step in gp130-mediated terminal differentiation and growth arrest of a myeloid cell line. Proc Natl Acad.Sci USA *93:* 3963–3966.

Nakajima, K., Y. Yamanaka, K. Nakae, H. Kojima, M. Ichiba, N. Kiuchi, T. Kitaoka, T. Fukada, M. Hibi, T. Hirano (1996) A central role for Stat3 in IL-6-induced regulation of growth and differentiation in M1 leukemia cells. EMBO J *15:* 3651–3658.

Neel, B.G., N.K. Tonks (1997) Protein tyrosine phosphatases in signal transduction. Curr Opin Cell Biol *9:* 193–204.

Nichols, J., I. Chambers, A. Smith (1994) Derivation of germline competent embryonic stem cells with combination of interleukin-6 and soluble interleukin-6 receptor. Exp Cell Res *215:* 237–239.

Nichols, J., D. Davidson, T. Taga, K. Yoshida, I. Chambers, A.G. Smith (1996) Complementary tissue-specific expression of LIF and LIF-receptor mRNAs in early mouse embryogenesis. Mech Dev *57:* 123–131.

Nichols, J., B. Zevnik, K. Anastassiadis, H. Niwa, D. Klewe-Nebenius, I. Chambers, H. Scholer, A. Smith (1998) Formation of pluripotent stem cells in the mammalian embryo depends on the POU transcription factor Oct4. Cell *95:* 379–391.

Niwa, H., T. Burdon, I. Chambers, A. Smith (1998) Self-renewal of pluripotent embryonic stem cells is mediated via activation of STAT3. Genes Dev *12:* 2048–2060.

Pages, G., P. Lenormand, G. L'Allemain, J.C. Chambard, S. Meloche, J. Pouyssegur (1993) Mitogen-activated protein kinases p42mapk and p44mapk are required for fibroblast proliferation. Proc Natl Acad Sci USA *90:* 8319–8323.

Pennica, D., K.J. Shaw, T.A. Swanson, M.W. Moore, D.L. Shelton, K.A. Zioncheck, A. Rosenthal, T. Taga, N.F. Paoni, W.I. Wood (1995) Cardiotrophin-1. Biological activities and binding to the leukemia inhibitory factor receptor/gp130 signaling complex. J Biol Chem *270:* 10915–10922.

Qu, C.K., G.S. Feng (1998) Shp-2 has a positive regulatory role in ES cell differentiation and proliferation. Oncogene *17:* 433–439.

Qu, C.K., Z.Q. Shi, R. Shen, F.Y. Tsai, S.H. Orkin, G.S. Feng (1997) A deletion mutation in the SH2-N domain of Shp-2 severely suppresses hematopoietic cell development. Mol Cell Biol *17:* 5499–5507.

Raabe, T., J. Riesgo-Escovar, X. Liu, B.S. Bausenwein, P. Deak, P. Maroy, E. Hafen (1996) DOS, a novel pleckstrin homology domain-containing protein required for signal transduction between sevenless and Ras1 in Drosophila. Cell *85:* 911–920.

Resnitzky, D., L. Hengst, S.I. Reed (1995) Cyclin A-associated kinase activity is rate limiting for entrance into S phase and is negatively regulated in G1 by p27Kip1. Mol Cell Biol *15:* 4347–4352.

Rodig, S.J., M.A. Meraz, J.M. White, P.A. Lampe, J.K. Riley, C.D. Arthur, K.L. King, K.C. Sheehan, L. Yin, D. Pennica, E.M. Johnson, Jr., R.D. Schreiber (1998) Disruption of the Jak1 gene demonstrates obligatory and nonredundant roles of the Jaks in cytokine-induced biologic responses. Cell *93:* 373–383.

Sasse, J., U. Hemmann, C. Schwartz, U. Schniertshauer, B. Heesel, C. Landgraf, J. Schneider-Mergener, P.C. Heinrich, F. Horn (1997) Mutational analysis of acute-phase response factor/Stat3 activation and dimerization. Mol Cell Biol *17:* 4677–4686.

Savatier, P., S. Huang, L. Szekely, K.G. Wiman, J. Samarut (1994) Contrasting patterns of retinoblastoma protein expression in mouse embryonic stem cells and embryonic fibroblasts. Oncogene *9:* 809–818.

Savatier, P., H. Lapillonne, L.A. van Grunsven, B.B. Rudkin, J. Samarut (1996) Withdrawal of differentiation inhibitory activity/leukemia inhibitory factor up-regulates D-type cyclins and cyclin-dependent kinase inhibitors in mouse embryonic stem cells. Oncogene *12:* 309–322.

Saxton, T.M., M. Henkemeyer, S. Gasca, R. Shen, D.J. Rossi, F. Shalaby, G.-S. Feng, T. Pawson (1997) Abnormal mesoderm patterning in mouse embryos mutant for the SH2 tyrosine phosphatase SHP-2. EMBO J *16:* 2352–2364.

Schaefer, T.S., L.K. Sanders, D. Nathans (1995) Cooperative transcriptional activity of Jun and Stat3 beta, a short form of Stat3. Proc Natl Acad Sci USA *92:* 9097–9101.

Schaefer, T.S. L.K. Sanders, O.K. Park, D. Nathans (1997) Functional differences between Stat3alpha and Stat3beta. Mol Cell Biol *17:* 5307–5316.

Servidei, T., Y. Aoki, S.E. Lewis, A. Symes, J.S. Fink, S.A. Reeves, (1998) Coordinate regulation of STAT signaling and c-fos expression by the tyrosine phosphatase SHP-2. J Biol Chem *273:* 6233–6241.

Sheng, Z., K. Knowlton, J. Chen, M. Hoshijima, J.H. Brown, K.R. Chien (1997) Cardiotrophin 1 (CT-1) inhibition of cardiac myocyte apoptosis via a mitogen-activated protein kinase-dependent pathway. Divergence from downstream CT-1 signals for myocardial cell hypertrophy. J Biol Chem *272:* 5783–5791.

Smith, A. (1998) Cell therapy: In search of pluripotency. Curr Biol *8:* 802–804.

Smith, A.G. (1992) Mouse embryo stem cells: Their identification, propagation and manipulation. Semin Cell Biology *3:* 385–399.

Smith, A.G., J.K. Heath, D.D. Donaldson, G.G. Wong, J. Moreau, M. Stahl, D. Rogers (1988) Inhibition of pluripotential embryonic stem cell differentiation by purified polypeptides. Nature *336:* 688–690.

Smith, A.G., M.L. Hooper (1987) Buffalo rat liver cells produce a diffusible activity which inhibits the differentiation of murine embryonal carcinoma and embryonic stem cells. Dev Biol *121:* 1–9.

Smith, A.J.H., M.A. De Sousa, B. Kwabi-Addo, A. Heppell-Parton, H. Impey, P. Rabbitts (1995) A site-directed chromosomal translocation induced in embryonic stem cells by Cre/*loxP* recombination. Nat Genet *9:* 376–385.

Stahl, N., T.G. Boulton, T. Farruggella, N.Y. Ip, S. Davis, B.A. Witthuhn, F.W. Quelle, O. Silvennoinen, G. Barbieri, S. Pellegrini, J.N. Ihle, G.D. Yancopoulos (1994) Association and activation of Jak-Tyk kinases by CNTF-LIF-OSM-IL-6 beta receptor components. Science *263:* 92–95.

Stahl, N., T.J. Farrugella, T.G. Boulton, Z. Zhong, J.E. Darnell, G.D. Yancopoulos (1995) Choice of STATs and other substrates specified by modular tyrosine-based motifs in cytokine receptors. Science *267:* 1349–1353.

Starr, R., U. Novak, T.A. Willson, M. Inglese, V. Murphy, W.S. Alexander, D. Metcalf, N.A. Nicola, D.J. Hilton, M. Ernst (1997) Distinct roles for leukemia inhibitory factor receptor α-chain and gp130 in cell type-specific signal transduction. J Biol Chem *272:* 19982–19986.

Stewart, C.L., P. Kaspar, L.J. Brunet, H. Bhatt, I. Gadi, F. Kontgen, S.J. Abbondanzo (1992) Blastocyst implantation depends on maternal expression of leukaemia inhibitory factor. Nature *359:* 76–79.

Symes, A., N. Stahl, S.A. Reeves, T. Farruggella, T. Servidei, T. Gearan, G. Yancopoulos, J.S. Fink (1997) The protein tyrosine phosphatase SHP-2 negatively regulates ciliary neurotrophic factor induction of gene expression. Curr Biol *7:* 697–700.

Taga, T., T. Kishimoto (1997) Gp130 and the interleukin-6 family of cytokines. Annu Rev Immunol *15:* 797–819.

Takahashi-Tezuka, M., Y. Yoshida, T. Fukada, T. Ohtani, Y. Yamanaka, K. Nishida, K. Nakajima, M. Hibi, T. Hirano (1998) Gab1 acts as an adapter molecule linking the cytokine receptor gp130 to ERK mitogen-activated protein kinase. Mol Cell Biol *18:* 4109–4117.

Takeda, K., K. Noguchi, W. Shi, T. Tanaka, M. Matsumoto, N. Yoshida, T. Kishimoto, S. Akira (1997) Targeted disruption of the mouse *Stat3* gene leads to early embryonic lethality. Proc Natl Acad Sci USA *94:* 3801–3804.

Tang, T.L., R.M. Freeman, Jr., A.M. O'Reilly, B.G. Neel, S.Y. Sokol (1995) The SH2-containing protein-tyrosine phosphatase SH-PTP2 is required upstream of MAP kinase for early Xenopus development. Cell *80:* 473–483.

Thomson, J. A., J. Itskovitz-Eldor, S.S. Shapiro, M.A. Waknitz, J.J. Swiergiel, V.S. Marshall, J.M. Jones (1998) Embryonic stem cell lines derived from human blastocysts. Science *282:* 1145–1147.

Thomson, J. A., V.S. Marshall (1998) Primate embryonic stem cells. Curr Top Dev Biol *38:* 133–165.

Turkson, J., T. Bowman, R. Garcia, E. Caldenhoven, R.P. De Groot, R. Jove (1998) Stat3 activation by Src induces specific gene regulation and is required for cell transformation. Mol Cell Biol *18:* 2545–2552.

Van Vactor, D., A.M. O'Reilly, B.G. Neel (1998) Genetic analysis of protein tyrosine phosphatases. Curr Opin Genet Dev *8:* 112–126.

Wen, Z., Z. Zhong, J.E. Darnell, Jr. (1995) Maximal activation of transcription by Stat1 and Stat3 requires both tyrosine and serine phosphorylation. Cell *82:* 241–250.

Whitman, M., D.A. Melton (1992) Involvement of p21ras in Xenopus mesoderm induction. Nature *357:* 252–254.

Wianny, F., F.X. Real, C.L. Mummery, M. Van Rooijen, J. Lahti, J. Samarut, P. Savatier (1998) G1-phase regulators, cyclin D1, cyclin D2, and cyclin D3: Up-regulation at gastrulation and dynamic expression during neurulation. Dev Dyn *212:* 49–62.

Williams, R.L., D.J. Hilton, S. Pease, T.A. Willson, C.L. Stewart, D.P. Gearing, E.F. Wagner, D. Metcalf, N.A. Nicola, N.M. Gough (1988) Myeloid leukaemia inhibitory factor maintains the developmental potential of embryonic stem cells. Nature *336:* 684–687.

Yamanaka, Y., K. Nakajima, T. Fukada, M. Hibi, T. Hirano (1996) Differentiation and growth arrest signals are generated through the cytoplasmic region of gp130 that is essential for STAT3 activation. EMBO J *15:* 1557–1565.

Yoshida, K., I. Chambers, J. Nichols, A. Smith, M. Saito, K. Yasukawa, M. Shoyab, T. Taga, T. Kishimoto (1994) Maintenance of the pluripotential phenotype of embryonic stem cells through direct activation of gp130 signalling pathways. Mech Dev *45:* 163–171.

Yoshida, K., T. Taga, M. Saito, S. Suematsu, A. Kumanogoh, T. Tanaka, H. Fujiwara, M. Hirata, T. Yamagami, T. Nakahata, T. Hirabayashi, Y. Yoneda, K. Tanaka, W.-Z. Wang, C. Mori, K. Shiota, N. Yoshida, T. Kishimoto (1996). Targeted disruption of gp130, a common signal transducer for the interleukin 6 family of cytokines, leads to myocardial and hematological disorders. Proc Natl Acad Sci USA *93:* 407–411.

Yu, C.L., D.J. Meyer, G.S. Campbell, A.C. Larner, C. Carter-Su, J. Schwartz, R. Jove (1995). Enhanced DNA-binding activity of a Stat3-related protein in cells transformed by the Src oncoprotein. Science *269:* 81–83.

Zhong, Z., Z. Wen, J.E. Darnell, Jr. (1994) Stat3: A STAT family member activated by tyrosine phosphorylation in response to epidermal growth factor and interleukin-6. Science *264:* 95–98.

Cells
Tissues
Organs

Cells Tissues Organs 1999;165:144–152

Accepted after revision: March 14, 1999

Oct-4: Lessons of Totipotency from Embryonic Stem Cells

Maurizio Pesce[a] Konstantinos Anastassiadis[a] Hans R. Schöler[a,b]

[a]EMBL, Gene Expression Programme, Heidelberg, Germany; [b]Center for Transgenesis and Germ Cell Research, School of Veterinary Medicine, New Bolton Center, University of Pennsylvania, Philadelphia, Pa., USA

Key Words

Oct-4 · Embryonic stem cell · Transcription factor · Germ cell

Abstract

The Oct-4 transcription factor is expressed in embryonic stem cells and germ cells of the mouse embryo. Cells that are driven into embryonic and extraembryonic somatic differentiation lose Oct-4 expression. Oct-4 is crucial for the maintenance of the totipotency of embryonic cells at early developmental stages as demonstrated recently by homologous recombination. Since during embryogenesis Oct-4 is restricted to germ cells, it is possible that it plays a key role for the germline lineage by preventing differentiation of these cells during gastrulation. Thus, Oct-4 appears to be a key regulator of totipotency during the life cycle of mammals.

Abbreviations used in this paper

DE	distal enhancer
EC	embryonic carcinoma cells
ES	embryonic stem cells
ICM	inner cell mass
PE	proximal enhancer
PGC	primordial germ cells
RARE	retinoic acid-responsive element

Introduction

In the mouse, the expression of Oct-4 is confined to cells that share a totipotent phenotype. This evidence has led to the idea that this transcription factor may cyclically 'trace' the totipotency of embryonic and germ cells through the lifecycle until the late phases of gametogenesis in the adult. The concept of the totipotent cycle (fig. 1) predicts that any totipotent cell losing Oct-4 expression would exit the cycle and undergo somatic differentiation.

Oocytes of invertebrates such as Drosophila or Caenorhabditis, and vertebrates such as anuran amphibia contain localized or gradually distributed factors that determine the fate of the cells arising during early cleavage stages. By contrast, fate determination of embryonic cells in mammalians is likely dictated by their position in the developing embryo. For example, outer cells of the compacted morula are segregated as trophoblastic cells by virtue of their polarization. The mechanism(s) causing trophoblast differentiation might depend on specific cell contacts that these polarized cells establish with each other. Since the inner compartment of the embryo is excluded from surface polarization of outer cells, inner cells might be prevented from differentiation. Alternatively, specific cell contacts established among inner cells might prevent their differentiation. Strikingly, the pattern of protein expression in outer cells of precavitation embryos diverges from the pattern of the inner cells [Handyside and Johnson, 1978]. This suggests that positionally regu-

KARGER

Fax + 41 61 306 12 34
E-Mail karger@karger.ch
www.karger.com

1422–6405/99/1654–0144$17.50/0

Accessible online at:
www.karger.com/journals/cto

Prof. Dr. Hans R. Schöler
EMBL, PO Box 102209
Meyerhoferstrasse 1
D–69012 Heidelberg (Germany)
Fax +49 6221 387518, E-Mail Schoeler@EMBL-Heidelberg.de

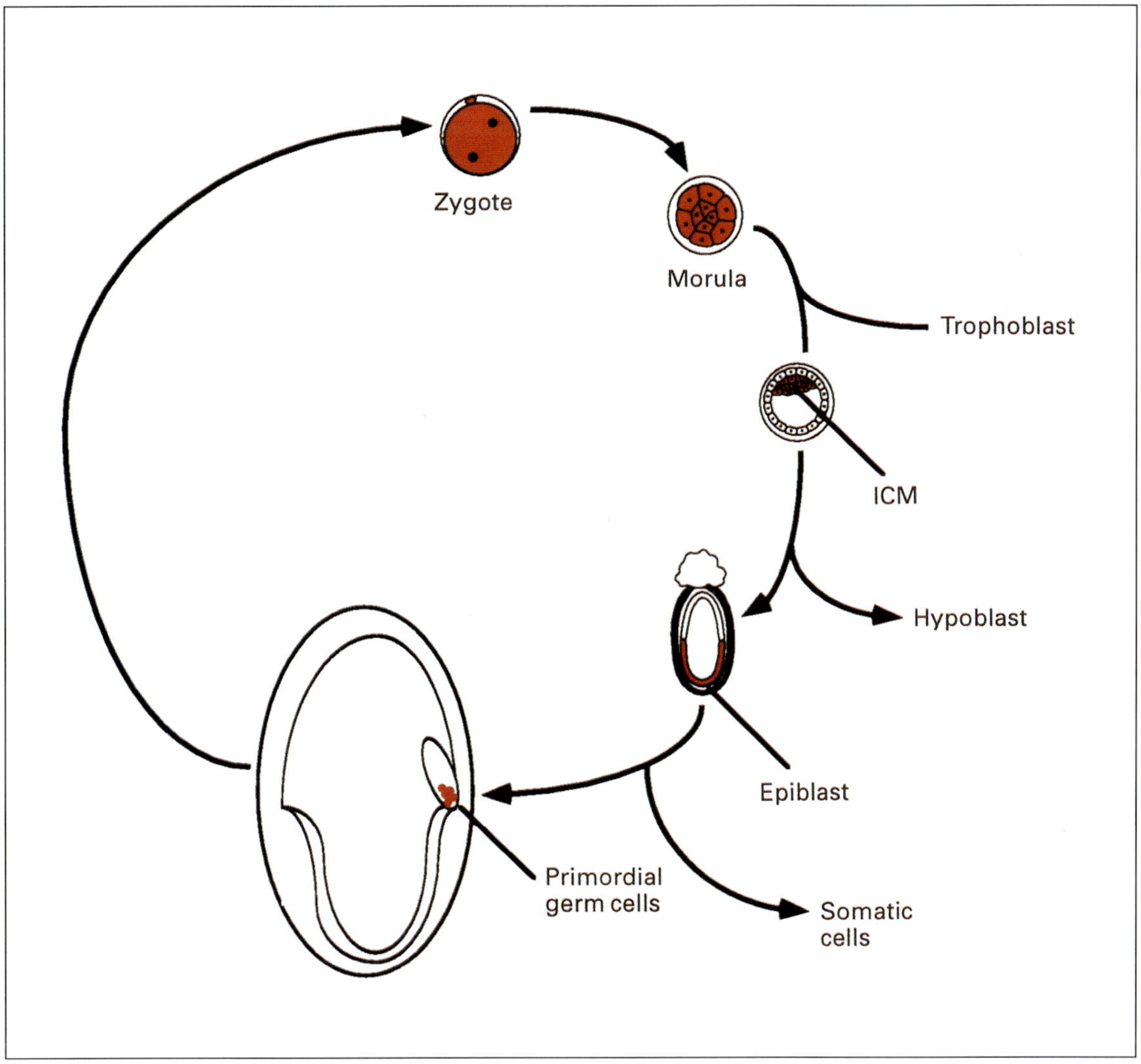

Fig. 1. Totipotent cycle in mammals. During early cleavage stages Oct-4 is expressed in all blastomeres. After differentiation of trophoblast cells, Oct-4 is restricted to cells of the ICM. After implantation, Oct-4 expression is maintained in epiblast (also termed primitive ectoderm) cells whereas it is downregulated in the extraembryonic hypoblast (also termed primitive endoderm). During gastrulation, Oct-4 is downregulated in an anterior to posterior manner in the somatic lineages and confined to primordial germ cells.

lated gene expression may cause fate determination of outer cells compared to inner cells before the morphological appearance of trophoblast and the inner cell mass (ICM). By contrast, it has recently been shown that a gradient of leptin/STAT3 proteins may be formed within the oocyte and among the blastomeres during early cleavage stages [Antczak and Van Blerkom, 1997]. This has been taken as an indication that a differential distribution of transcription factors may occur in the mouse oocyte and establish a differential gene expression pattern in inner cells compared to outer cells in the developing embryo.

The segregation of the germline in mouse embryos takes place in an extraembryonic region at the base of the allantoic bud where cell contacts, along with local signals, may induce differentiation of some epiblast cells to primordial germ cells (PGCs). It has been shown that, prior to gastrulation, epiblast cells have the potency to develop to germ cells if allocated to this extraembryonic position [Tam and Zhou, 1996]. This suggests that totipotent cells are not committed to become germ cells by inheritance of 'determinants' in a classical sense but are likely committed by local signals, which maintain their totipotency and drive their lineage to germ cells.

Many clues suggest that Oct-4 is regulated positionally in the developing embryo. In this view, Oct-4 expression can be considered as a part of a mechanism, which protects embryonic and germ cells from somatic fate during embryonic development. Thus, the 'decision' that leads to Oct-4 expression or downregulation in a given totipotent cell might be one of the first events causing embryonic cells to maintain a totipotent state or to be triggered into differentiation.

The recent abolition of Oct-4 function in vivo by homologous recombination has demonstrated that Oct-4-deficient embryos die shortly after implantation due to a lack of formation of ICM. This demonstrates that the Oct-4 transcription factor plays a primary role in the development of the early embryo and the regulation of the totipotent state of embryonic cells.

Embryonic stem (ES) and embryonic carcinoma (EC) cells are derived from cells of the ICM and epiblast, respectively. With respect to gene expression patterns and the potential to differentiate, these cells are very similar to totipotent embryonic cells. ES cells are considered totipotent since they contribute to all embryonic lineages, including the germline of chimeric embryos. Due to this feature, these cells are used for the generation of mice carrying germline mutations of genes whose functions can then be studied in vivo. ES and EC cells express Oct-4 and have been proven to be a suitable system with which to study its function and regulation.

Oct-4: A Transcription Factor Containing a POU Domain

POU transcription factors contain a bipartite DNA binding domain called the POU domain. Most POU transcription factors bind to the DNA sequence, called the octamer motif, originally found as a *cis*-acting element important for the immunoglobulin enhancer activity [Herr and Cleary, 1995]. These factors have been isolated from a variety of different vertebrates and invertebrates and have been grouped into six different classes according to their sequence similarity [Ryan and Rosenfeld, 1997].

Oct-4 is a member of the class V of POU transcription factors. It was initially identified as a DNA binding activity in extracts from ES and EC cells [Lenardo et al., 1989; Schöler et al., 1989]. The *Oct-4* gene encodes a protein of 352 amino acids [Schöler et al., 1990; Okazawa et al., 1991] and is organized in five exons [Yeom et al., 1991]. In mammals *Oct-4* orthologs have been identified in human [Takeda et al., 1992] and bovine. Mammalian *Oct-4* genes share high structural homology and conservation of aminoacid sequence suggesting that they may function in a similar fashion. Interestingly, other members of the class V of POU transcription factors are expressed in the oocyte as maternal transcripts or in undifferentiated cells during cleavage stages. This raises the possibility that members of the class V of POU transcription factors may have evolved in different species as key regulators of early embryonic development [Ryan and Rosenfeld, 1997].

Oct-4 activates transcription from binding sites located proximally or distantly from the transcriptional start site. However, it can activate transcription in a distance-independent manner only in undifferentiated ES and EC cells which express coactivators bridging the remotely bound Oct-4 to the trancriptional machinery. In differentiated cells, the adenoviral protein E1A that functions as an Oct-4 coactivator mimics this activity [Schöler et al., 1991]. Direct contacts between Oct-4 POU domain and the constant region 3 of E1A are involved in the functional cooperativity between the two factors [Schöler et al., 1991]. Recently, new biochemical analyses performed in our lab showed that the Oct-4/E1A interaction is also mediated by other domains of E1A. In particular it has been found that the CR1 of E1A mediates the interaction with Oct-4 [Brehm et al., 1999]. The activity of the C-terminal transactivation domain of Oct-4 may be modulated cell-specifically by phosphorylation suggesting that Oct-4 is a target of intracellular signaling cascades [Brehm et al., 1997].

Enhancer and promoter sites activated or repressed by Oct-4 have been identified. An enhancer element activated by Oct-4 has been found in the 3′-UTR of the *FGF-4* gene [Curatola and Basilico, 1990]. Recent analyses showed that this element is activated by Oct-4 in cooperation with the stem cell-specific HMG box transcription factor Sox-2 [Yuan et al., 1995]. Sox-2 interacts and functionally synergizes with Oct-4 [Ambrosetti et al., 1997]. Another *cis*-acting element bound and activated by Oct-4 was recently found within the first inton of the *Osteopontin (Opn)* gene, encoding an extracellular matrix component that is supposed to participate in the early events of primitive endodermal cell migration [Botquin et al., 1998]. The *Opn* enhancer contains a DNA sequence to which Oct-4 binds and activates transcription as a dimer.

Interestingly, a Sox-2 site has been found in close proximity to the Oct-4 binding site also within the *Opn* intron. In this case, Sox-2 seems to repress the Oct-4 mediated transactivation of *Opn* gene possibly interfering with Oct-4 dimer formation [Botquin et al., 1998].

The *Rex-1* promoter also contains a regulatory element activated by Oct-4 in pluripotent cells. Rex-1 is a zinc finger transcription factor expressed specifically in undifferentiated stem cells [Hosler et al., 1993]. Oct-4 activates transcription of the *Rex-1* gene in cooperation with a novel, stem cell-specific DNA binding activity named Rox-1 [Hosler et al., 1993; Ben-Shushan et al., 1998].

Oct-4 functions not just as an activator of genes expressed in stem cells, but may also prevent the expression of genes that are activated during the stem cell differentiation pathway. Oct-4 repression has been discussed for the human chorionic gonadotropin α and β gene promoters in choriocarcinoma cells [Liu et al., 1996, 1997]. Since both hormones are expressed by trophoblastic cells as they differentiate at the surface of the compacted morula, it is likely that downregulation of Oct-4 in outer cells of the morula may cause the activation of genes involved in the trophoblast lineage. This notion is supported by the analysis of Oct-4-deficient embryos (see below).

Oct-4: A Tracer of the Totipotency of Embryonic Cells

Present as a maternally inherited transcript, embryonic Oct-4 expression starts at the 8-cell stage when a high amount of Oct-4 protein is found in the nuclei of blastomeres [Palmieri et al., 1994]. Until compaction of the morula, there is no cellular restriction of Oct-4 expression. However, after cavitation Oct-4 is only maintained in cells that will give rise to the ICM of the blastocyst (fig. 1), while it is downregulated in the two other tissues, the trophectoderm and the hypoblast (also termed primitive endoderm), that will give rise to chorion and visceral/parietal endoderm, respectively [Yeom et al., 1991; Palmieri et al., 1994]. Thereafter, Oct-4 is restricted to the primitive ectoderm (also termed epiblast) to be finally downregulated during gastrulation in an anterior-posterior manner [Yeom et al., 1996]. The only cells that still maintain Oct-4 expression after the completion of gastrulation are PGCs which are segregated at the midgastrula stage in an extraembryonic position at the base of the allantoic bud [for a review, see Pesce et al., 1998a]. These cells are the precursors of the gametes of the adult that after fertilization will give rise to a new organism (fig. 1).

Treatment of ES and EC cells with retinoic acid, DMSO or other differentiating agents causes a rapid downregulation of *Oct-4* [Schöler et al., 1990a, b; Minucci et al., 1996]. In an experiment in which EC cells were fused with fibroblasts *Oct-4* expression was extinguished in hybrid cells. These hybrids displayed a neuronal phenotype [Ben-Shushan et al., 1993; Shimazaki et al., 1993]. However, when neuronal-like hybrid cells were transfected with both Oct-4 and E1A expression vectors, they lost the neuronal phenotype and reverted to a cell type that could again be induced to differentiate [Shimazaki et al., 1993]. Thus, the Oct-4 transactivating function correlates with the activation of genes involved in the maintenance of the totipotency of stem cells. On the other hand, since in some cases Oct-4 has been found to repress transcription in differentiated cells (see above), it is possible that it may also repress, directly or indirectly, genes whose expression would lead to somatic fate determination of totipotent cells. In this scenario Oct-4 would set a stem cell-specific gene expression pattern that would prevent the differentiation of stem cells.

Recently, a homologous recombination experiment that created an *Oct-4* null allele in ES cells has been performed [Nichols et al., 1998]. Embryos homozygous for this *Oct-4* mutation were able to form blastocysts at 3.5 days of gestation, but they were reabsorbed shortly after implantation. In vitro culture of Oct-4$^{-/-}$ morulae showed that in the absence of Oct-4, the proliferation/survival of blastomeres is normal until the cavitation stage (fig. 2). However, in the absence of Oct-4, cells that at this point would normally give rise to the ICM were driven into trophoblast differentiation. Furthermore, inner cells immunosurgically isolated from Oct-4-deficient morulae also differentiated to trophoblastic cells demonstrating that, in the absence of Oct-4, the totipotency of the inner cells was abolished. Mutant embryos showed a reduced expression of FGF-4 (fig. 3; see above). Addition of FGF-4 to the culture medium did not rescue the failure of ICM formation ind Oct-4$^{-/-}$ embryos, but improved the production of extraembryonic ectodermal cells. This is in agreement with recent results showing that, whilst required for the maintenance of the ICM in the embryo possibly stimulating the growth of stem cells in an autocrine manner [Feldman et al., 1994], FGF-4 may stimulate survival and/or proliferation of cells arising from the ICM in a paracrine manner [Wilder et al., 1997]. Altogether these results demonstrate that Oct-4 is required for preventing somatic differentiation of ICM cells as the first somatic lineage

a

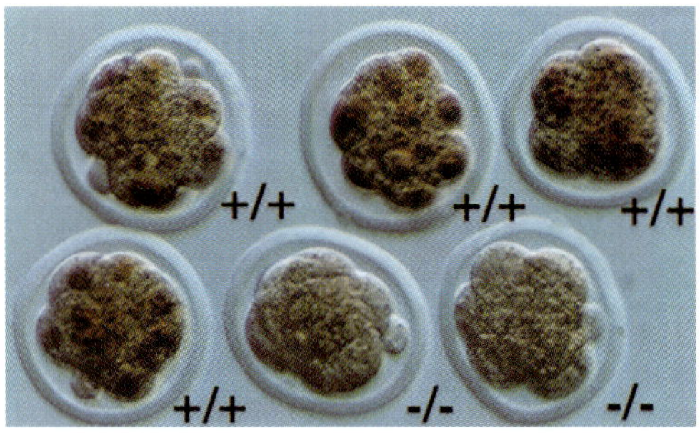

b

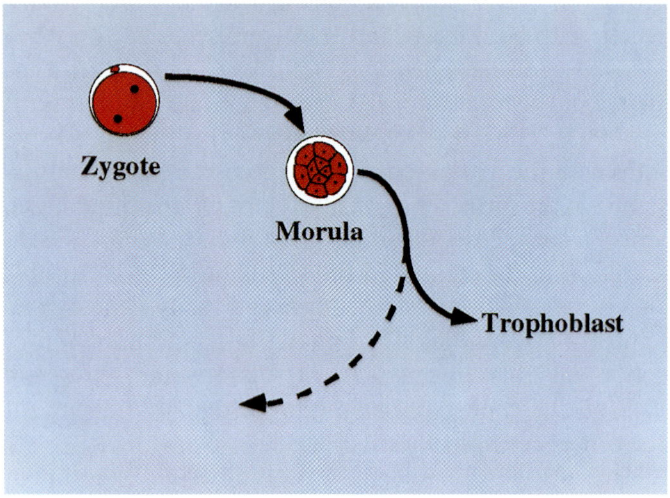

Fig. 2. Lack of Oct-4 function does not prevent cleavage and morula formation. **a** Whole mount immunohistochemistry of wt (+/+) and *Oct-4* null (–/–) embryos. The size and the number of cells are not altered in the mutant embryos, showing that Oct-4 loss of function does not impair development prior to cavitation. **b** Lack of Oct-4 causes failure of ICM formation. In the absence of Oct-4 all embryonic cells are fated to trophoblast differentiation and embryos do not progress any further in development [Nichols et al., 1998]. This shows that Oct-4 loss of function causes embryonic cells to exit the totipotent cycle when the first lineage decision occurs in the embryo.

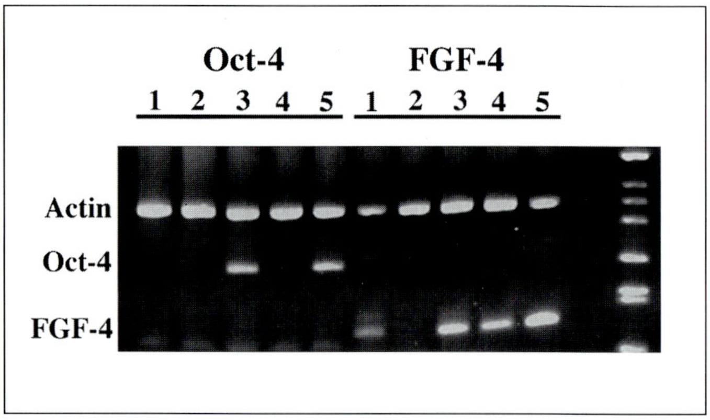

Fig. 3. Expression of FGF-4 in individual wt and Oct-4$^{-/-}$ embryos as detected by RT-PCR. Lack of Oct-4 causes a partial reduction of FGF-4 expression in embryos 1 and 4 and a complete lack of FGF-4 in embryo 2. This is in agreement with results which showed that *FGF-4* is regulated by an enhancer activated by Oct-4 [Ambrosetti et al., 1997]. Preparation of RNA and RT-PCR was done as described in Nichols et al. [1998].

(trophoblast) splits from the stem cell lineage (fig. 2). Thus, at least at this stage of embryonic development, Oct-4 fulfills the requirements of a key regulator for the totipotency of embryonic cells, as the model of the totipotent cycle predicts (fig. 1).

Regulation of Oct-4 in Totipotent Cells

Oct-4 expression depends on a TATA-less promoter [Okazawa et al., 1991] and two enhancer elements (fig. 4). A GC box located about 30 bp upstream of the major transcriptional start site mediates Oct-4 basal promoter activity [Sylvester and Schöler, 1994]. This site is occupied by transcription factors in vivo [Minucci et al., 1996] and represents a binding site with high affinity for Sp1 and Sp3 transcription factors (fig. 5). Recent experiments performed in *Sp1* and *Sp3* null ES cells suggested that Sp1 and Sp3 may interchangeably activate the *Oct-4* promoter [Pesce, unpubl. data]. The GC box is partially overlapping with a retinoic acid-responsive element (RARE) composed of three direct repeats to which members of the nuclear receptor family can bind [Sylvester and Schöler, 1994]. this RARE likely represents a negative element involved in retinoic acid-induced *Oct-4* downregulation. Recently, a novel *Oct-4* RARE binding complex has been found in extracts from EC cells that had been treated with retinoic acid [Fuhrmann, unpubl. data]. This complex is likely involved in early *Oct-4* downregulation in these cells.

Analysis of transgenic mice carrying LacZ reporter constructs in which different regions of the *Oct-4* gene were deleted provided evidence of two distinct *Oct-4* enhancers. These two elements are located in a 5-kb region upstream of the basal promoter and called the distal enhancer (DE) and the proximal enhancer (PE) due to their positions with respect to the promoter (fig. 4). They are essential for the cell-specific expression of *Oct-4* and their activity has a reciprocal pattern during embryogenesis [Okazawa et al., 1991; Yeom et al., 1996]. The DE acti-

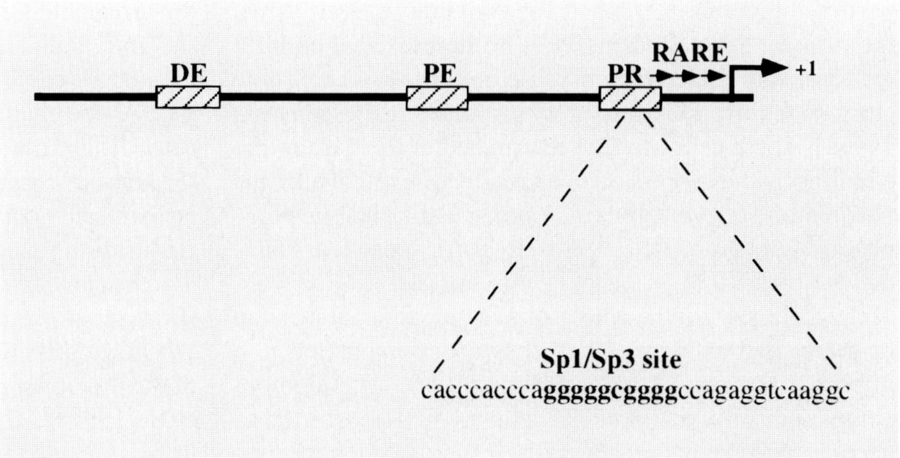

Fig. 4. Structure of the *Oct-4* upstream regulatory region. The basal promoter (PR) including the Sp1/Sp3 site and the RARE is highlighted. Two elements, the PE and DE which drive *Oct-4* cell-specific activity [Yeom et al., 1996], are also shown.

vates *Oct-4* in morula, ICM and PGCs, while the PE is necessary for *Oct-4* expression in the epiblast [Yeom et al., 1996]. Strikingly, the isolated DE and PE elements drive a reporter gene expression specifically in ES and EC cells, respectively, suggesting that these cells may express specific sets of transcription factors activating *Oct-4* enhancers in a cell-specific fashion [Yeom et al., 1996]. In vivo footprinting experiments identified sequences in both enhancer elements that are protected by binding of transcription factors in both cell types [Minucci et al., 1996]. Thus, there is an apparent contradiction between cell-specific activity and binding site occupancy of DE and PE. This suggests that other *cis*-acting elements within the enhancers and other regions of the *Oct-4* locus play a role in the cell-type specificity of DE and PE.

Treatment of EC and ES cells with retinoic acid leads to rapid loss of in vivo occupancy on the two enhancer elements and the promoter by transcription factors [Minucci et al., 1996]. This loss is paralleled by *Oct-4* gene downregulation suggesting that differentiation is associated with changes of the chromatin structure leading to *Oct-4* transcriptional repression.

Oct-4 and the Germline

At the end of gastrulation, *Oct-4* expression is confined to PGCs [Yeom et al., 1996]. PGCs arise in a posterior region of the embryo from progenitor cells in the extraembryonic mesodermal compartment, located at the base of the allantois (fig. 1). Prior to gastrulation, these precursors reside within a wide area of the epiblast juxtaposed to the extraembryonic ectoderm [Lawson and Hage, 1994]. This

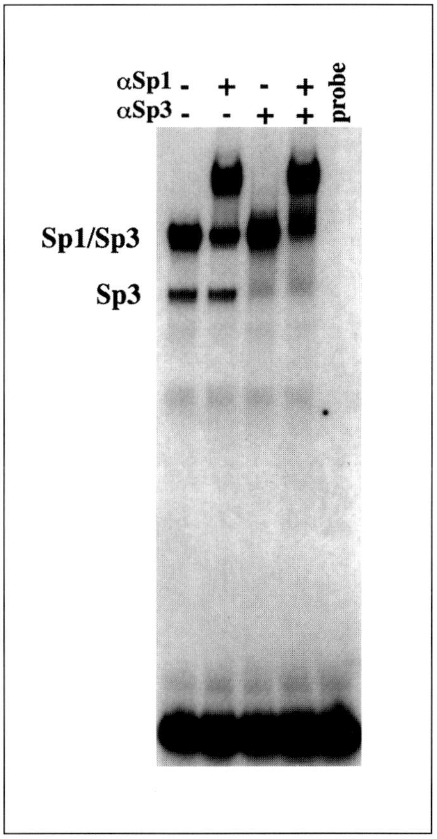

Fig. 5. GC box within the *Oct-4* promoter is a high affinity-binding site for Sp1 and Sp3. The major complexes formed by incubating whole cell extracts from ES cells with a ^{32}P-labeled probe encompassing the GC box are formed by Sp1 and Sp3, as assayed by the use of specific Sp1 and Sp3 antibodies. Sp3 complexes and Sp1 complex are supershifted by using a combination of both antibodies. Functional studies in Sp1 and Sp3 deficient cells suggest that both transcription factors cooperate in activating *Oct-4* promoter in totipotent cells [Pesce, unpubl. data].

part of the epiblast is called the proximal region. When the primitive streak forms, cells of the proximal epiblast are carried to the extraembryonic mesoderm where some clusters acquire PGC features [discussed in Pesce et al., 1998a]. These cells are first recognized at the base of the allantois by the expression of a specific isoform of alkaline phosphatase enzyme, the tissue nonspecific alkaline phosphatase [Ginsburg et al., 1990]. Recent experiments have demonstrated that, prior to the onset of gastrulation, cells of the distal part of the epiblast have the same potential to be segregated as germ cells if they are transplanted to a proximal position [Tam and Zhou, 1996]. This demonstrates that proximal epiblast cells are not committed to become germ cells until they reach their extraembryonic location. Inductive signals may drive some clusters of pluripotent cells to become PGCs in this extraembryonic position. The 'switch' in the activity from the *Oct-4* proximal enhancer to the distal enhancer temporally and spatially coincides with the allocation of PGCs [Yeom et al., 1996], suggesting that the 'reprogramming' of the *Oct-4* expression occurring in germ cells may be under the control of the same signals. At the moment it is not clear how this switch is regulated.

It has been suggested that the allocation of the germ cells in an 'enclave' may enable them to avoid processes driving somatic cell differentiation during the specification of the body plan [Dixon, 1994]. In this view, the segregation of Oct-4-expressing cells in the extraembryonic mesoderm of mouse embryos might reflect the necessity for germline competent cells to escape from differentiative signals and maintain a totipotent phenotype.

Strikingly, during gastrulation epiblast cells undergo an extensive de novo methylation [reviewed in Jaenisch, 1997]. By contrast, the PGC genome is kept undermethylated [Monk et al., 1987] and undergoes erasure of the parental imprint due to a potent demethylation activity [reviewed in Surani, 1998]. *Oct-4* extinction in hybrid cells resulting from fusion of EC cells with fibroblasts parallels a rapid methylation of *Oct-4* promoter/enhancer regions [Ben-Shushan et al., 1993]. Thus, it is likely that epiblastic downregulation of *Oct-4* may be mediated by epigenetic modifications of the chromatin occurring in embryonic cells during gastrulation. Again, the extraembryonic localization of PGCs may protect them from methylation and *Oct-4* downregulation.

Oct-4 expression is maintained in germ cells until the onset of the sexual differentiation. In a recent study it was found that Oct-4 is expressed in germ cells of both sexes until the beginning of maturation stages of oogenesis and spermatogenesis [Pesce et al., 1998b]. In oocytes Oct-4 is downregulated coincidently with entry into prophase of the first meiotic division. In male germ cells it is expressed by the most undifferentiated spermatogonia (the type A spermatogonia), and is downregulated as these cells differentiate to proliferating B spermatogonia and the subsequent maturation stages. Since Oct-4 downregulation occurs coincidentally with the beginning of gamete maturation processes in both sexes, it is possible that Oct-4 absence is crucial to switch the germ cells into meiosis [Pesce et al., 1998a].

Oocytes express Oct-4 protein again after completion of the meiotic prophase I until the fully grown stage [Pesce et al., 1998b]. Thus, Oct-4 may have additional functions in oogenesis by controlling genes involved in the growth and acquisition of the competence to resume meiosis.

Conclusion

The results reviewed in this paper strongly support the notion that Oct-4 is a key regulator of totipotency in mammalian embryonic cells. However, it remains unclear how totipotency is accomplished. It is tempting to speculate that Oct-4 defines totipotency by setting a specific pattern of gene expression in embryonic cells, thus acting as a true determinant of their totipotency. The crucial role of Oct-4 for embryo development demonstrates that this is the case during the stages until implantation.

Since totipotency and germline competence in mammals do not imply inheritance of determinants such as those described in other animal models like *Caenorhabditis elegans*, Drosophila and anuran amphibia, Oct-4 cannot be considered as a determinant in a classical sense. Indeed, it is restricted to cells that are fated to a totipotent phenotype depending on their position in the embryo. Thus, it appears extremely important to define the array of the *cis*- and *trans*-acting factors, which regulate the embryonic pattern of Oct-4 expression and controlling its spatial restriction to the different cell types belonging to the totipotent cycle.

Another intriguing feature of Oct-4 that should be addressed is the role played by this transcription factor in germ cell determination. In particular, it would be of interest to test if a loss of Oct-4 causes a failure of PGCs to form. Indeed, if this transcription factor fulfills the role of a guardian of the totipotency of the germ cells, it is likely that it is absolutely necessary to maintain the germline during gastrulation. Due to the early lethality of embryos lacking Oct-4, this hypothesis must be addressed through a conditional homologous recombination approach. This will allow the selective abolishment of *Oct-4* expression at the midgastrula stage when the germ cell lineage is established in the extraembryonic mesoderm.

If the lineage of the germ cells is determined by signals restricted to the extraembryonic mesoderm, it is likely that *Oct-4* expression is required for the maintenance of germ cell totipotency but that it is not sufficient to determine the lineage of these cells. Thus, testing if ectopic Oct-4 expression is sufficient to produce germ cells from any totipotent cell irrespective of its location within the epiblast will prove whether Oct-4 acts as the sole determinant of the germline in mammals.

Acknowledgements

The authors are indebted to Dr. Catherine Ovitt for critically reading the manuscript. M.P. was supported by a European Community Biotechnology fellowship.

References

Ambrosetti, D.-C., C. Basilico, L. Dailey (1977) Synergistic activation of the fibroblast growth factor 4 enhancer by Sox-2 and Oct-3 depends on protein-protein interactions facilitated by a specific spatial arrangement of factor binding site. Mol Cell Biol *17:* 6321–6329.

Antczak, M., J. Van Blerkom (1997) Oocyte influences on early development: The regulatory proteins leptin and STAT3 are polarized in mouse and human oocytes and differentially distributed within the cells of the preimplantation stage embryo. Mol Hum Reprod *3:* 1067–1086.

Ben-Shushan, E., E. Pikarsky, A. Klar, Y. Bergman (1993) Extinction of Oct-3/4 gene expression in embryonal carcinoma × fibroblast somatic cell hybrids is accompanied by changes in the methylation status, chromatin structure and transcriptional activity of the Oct-3/4 upstream region. Mol Cell Biol *13:* 891–901.

Ben-Sushan, E., J.R. Thompson, L.J. Gudas, Y. Berman (1998) Rex-1, a gene encoding a transcription factor expressed in the early embryo, is regulated via Oct-3/4 and Oct-6 binding to an octamer site and a novel protein, Rox-1, binding to an adjacent site. Mol Cell Biol *18:* 1866–1878.

Botquin, V., H. Hess, G. Fuhrmann, K. Anastassiadis, M.K. Gross, G. Vriend, H.R. Schöler (1998) New POU dimer configuration mediates antagonistic control of an osteopontin preimplantation enhancer by Oct-4 and Sox-2. Genes Dev *12:* 2073–2090.

Brehm, A., K. Ohbo, H.R. Schöler (1997) The carboxy-terminal transactivation domain of Oct-4 acquires cell specificity through the POU domain. Mol Cell Biol *17:* 154–162.

Brehm, A., K. Ohbo, W. Zwerschke, V. Botquin, P. Jansen-Dürr, H.R. Schöler (1999) Synergism with germ line transcription factor Oct-4: Viral oncoproteins share the ability to mimic a stem cell-specific activity. Mol Cell Biol *19:* 2635–2643.

Curatola, A.M., C. Basilico (1990) Expression of the K-fgf proto-oncogene is controlled by 3′ regulatory elements which are specific for embryonal carcinoma cells. Mol Cell Biol *10:* 2475–2484.

Dixon, K.E. (1994) Evolutionary aspects of primordial germ cells formation; in Marsh, J., J. Goode (eds): Germline Development. Chichester, Wiley, pp 92–120.

Feldman, B., W. Poueymirou, V.E. Papaioannou, T.M. DeChiara, M. Goldfarb (1994) Requirement of FGF-4 for postimplantation mouse development. Science *267:* 246–249.

Ginsburg, M., M.H.L. Snow, A. McLaren (1990) Primordial germ cells in the mouse embryo during gastrulation. Development *110:* 521–528.

Handyside, A.H., M.H. Johnson (1978) Temporal and spatial patterns of the synthesis of tissue-specific polypeptides in the preimplantation mouse embryo. J Embryol Exp Morphol *44:* 191–199.

Herr, W., M.A. Cleary (1995) The POU domain: Versatility in transcriptional regulation by a flexible two-in-one DNA-binding domain. Genes Dev *9:* 1679–1693.

Hosler, B., M.B. Rogers, C.A. Kozak, L.J. Gudas (1993) An octamer motif contributes to the expression of the retinoic acid-regulated zinc finger gene Rex-1 (Zfp-42) in F9 teratocarcinoma cells. Mol Cell Biol *13:* 2919–2928.

Jaenisch, R. (1997) DNA methylation and imprinting: Why bother? Trends Genet *13:* 323–329.

Lawson, K.A., W.J. Hage (1994) Clonal analysis of the origin of primordial germ cells in the mouse; in Marsh, J., J. Goode (eds): Germline Development. Chichester, Wiley, pp 68–91.

Lenardo, M.J., L. Staudt, P. Robbins, A. Kuang, R.C. Mulligan, D. Baltimore (1989) Repression of the IgH enhancer in teratocarcinoma cells associated with a novel octamer factor. Science *243:* 544–546.

Liu, L., D. Laeman, M. Villalta, R.M. Roberts (1997) Silencing of the gene for the α-subunit of human chorionic gonadotropin by the embryonic transcription factor Oct-3/4. Mol Endocrinol *11:* 1651–1658.

Liu, L., R.M. Roberts (1996) Silencing of the gene for the β subunit of human chorionic gonadotropin by the embryonic transcription factor Oct-4. J Biol Chem *271:* 16683–16689.

Minucci, S., V. Botquin, Y.I. Yeom, A. Dey, I. Sylvester, D. Zand, K. Ozato, H.R. Schöler (1996) Retinoic acid-mediated down-regulation of Oct 3/4 coincides with the loss of promoter occupancy in vivo. EMBO J *15:* 888–899.

Monk, M., M. Boubelik, S. Lehnert (1987) Temporal and spatial changes in DNA methylation in the embryonic, extraembryonic and germ cell lineages during mouse embryo development. Development *99:* 371–382.

Nichols, J., B. Zevnik, K. Anastassiadis, H. Niwa, D. Klewe-Nebenius, I. Chambers, H. Schöler, A. Smith (1998) Formation of pluripotent stem cells in the mammalian embryo depends on the POU transcription factor Oct-4. Cell *95:* 379–391.

Okazawa, H., K. Okamoto, F. Ishino, T. Ishino-Kaneko, Y. Toyoda, M. Muramatsu, M.H. Hamada (1991) The oct3 gene, a gene for a embryonic transcription factor, is controlled by a retinoic acid repressible enhancer. EMBO J *10:* 2997–3005.

Palmieri, S., W. Peter, H. Hess, H.R. Schöler (1994) Oct-4 transcription factor is differentially expressed in the mouse embryo during establishment of the first two extraembryonic cell lineages involved in implantation. Dev Biol *166:* 259–267.

Pesce, M., M.K. Gross, H.R. Schöler (1998a) In line with our ancestors: Oct-4 and the mammalian germ. Bioessays *20:* 722–732.

Pesce, M., X. Wang, D. Wolgemuth, H.R. Schöler (1998b) Differential expression of the Oct-4 transcription factor during mouse germ cell differentiation. Mech Dev *71:* 89–98.

Ryan, A.K., M.G. Rosenfel (1997) POU domain family values: Flexibility, partnerships, and developmental codes. Genes Dev *11:* 1207–1225.

Shimazaki, T., H. Okazawa, H. Fuji, M. Ikeda, K. Tamai, R.D.G. McKay, M. Muramatsu, H. Hamada (1993) Hybrid cell extinction and re-expression of Oct-3 function correlates with differentiation potential. EMBO J *12:* 4489–4498.

Schöler, H.R., R. Balling, A.K. Hatzopoulos, N. Suzuki, P. Gruss (1989) Octamer binding proteins confer transcriptional activity in early mouse embryogenesis. EMBO J *8:* 2551–2557.

Schöler, H.R., T. Ciesiolka, P. Gruss (1991) A nexus between Oct-4 and E1A: Implication for gene regulation in embryonic stem cells. Cell *66:* 291–304.

Schöler, H.R., G.R. Dressler, R. Balling, H. Rohdewold, P. Gruss (1990b) Oct-4: A germline-specific transcription factor mapping to the mouse t-complex. EMBO J *7:* 2185–2195.

Schöler, H.R., S. Ruppert, N. Suzuki, K. Chowdhury, P. Gruss (1990a) New type of POU domain in germ line-specific protein Oct-4. Nature *344:* 435–439.

Surani, M.A. (1998) Imprinting and the initiation of gene silencing in the germ line. Cell *93:* 309–312.

Sylvester, I., H.R. Schöler (1994) Regulation of the *Oct-4* gene by nuclear receptors. Nucleic Acids Res *22:* 901–911.

Takeda, J., S. Seino, G.I. Bell (1992) Human Oct-3 gene family: cDNA sequences, alternative splicing, gene organization, chromosomal localization, and expression at low levels in adult tissues. Nucleic Acids Res *20:* 4613–4620.

Tam, P.P.L., S.X. Zhou (1996) The allocation of epiblast cells to ectodermal and germ-line lineages is influenced by the position of the cells in the gastrulating mouse embryo. Dev Biol *178:* 124–132.

Wilder, P.J., D. Kelly, K. Brigman, C.L. Peterson, T. Nowling, Q.-S. Gao, R.D. McComb, M.R. Capecchi, A. Rizzino (1997) Inactivation of the FGF-4 gene in embryonic stem cells alters the growth and/or the survival of their early differentiated progeny. Dev Biol *192:* 614–629.

Yuan, H.N., N. Corbi, C. Basilico, L. Dailey (1995) Developmental-specific activity of the FGF-4 enhancer requires the synergistic action of Sox-2 and Oct-3. Genes Dev *9:* 2635–2645.

Yeom, Y.I., G. Fuhrmann, C.E. Ovitt, A. Brehm, K. Ohbo, M. Gross, K. Hübner, H.R. Schöler (1996) Germline regulatory element of Oct-4 specific for the totipotent cycle of embryonal cells. Development *122:* 881–894.

Yeom, Y.I., H.-S. Ha, R. Balling, H.R. Schöler, K. Artzt (1991) Structure, expression and chromosomal localization of the Oct-4 gene. Mech Dev *35:* 171–179.

Cells
Tissues
Organs

Cells Tissues Organs 1999;165:153–164

Accepted after revision: August 18, 1999

Establishment of Ionic Channels and Signalling Cascades in the Embryonic Stem Cell-Derived Primitive Endoderm and Cardiovascular System

J. Hescheler[a] B.K. Fleischmann[a] M. Wartenberg[a] W. Bloch[b] E. Kolossov[a] G. Ji[a] K. Addicks[b] H. Sauer[a]

[a]Institute of Neurophysiology and [b]Institute of Anatomy, University of Cologne, Germany

Key Words

Embryonic stem cells · Embryoid body · Primitive endoderm · Angiogenesis · Cardiomyogenesis

Abstract

The first organ system to be established in early embryogenesis is the cardiovascular system which develops upon interaction with hypoblastic cells of the primitive endoderm. Here we focus on recent work on embryoid bodies derived from pluripotent embryonic stem (ES) cells. Ca^{2+} oscillations and Ca^{2+} signalling pathways during the differentiation of primitive endodermal cell layers are reported. Furthermore, the development-dependent expression of ion channels and the buildup of signalling cascades involved in the modulation of voltage-dependent L-type Ca^{2+} channels during early cardiomyogenesis and the formation of functional vascular structures in the process of vasculogenesis and angiogenesis are reviewed. We also report on the use of green fluorescent protein reporter gene expression under the control of cardiac-specific promoters, e.g. the human cardiac α-actin promoter, which enables the identification and in vivo characterization of cardiomyocytes at very early stages of cardiomyogenesis.

Introduction

Embryonic stem (ES) cells provide a unique cellular system to study commitment and differentiation of tissues under in vitro conditions [Gossler et al., 1986]. Cultured within embryoid bodies [Wobus et al., 1991], ES cells recapitulate among others the development of cardiomyocytes from very early (e.g. cardiac precursor cells)

Abbreviations used in this paper

BCECF	2′,7′-bis-(2-carboxyethyl)-5-(and-6)-carboxyfluorescein
CCh	carbachol
DiO	3,3′-dioctadecyloxacarbocyanine perchlorate
EDS	early developmental stage
ES cells	embryonic stem cells
GFP	green fluorescent protein
I_f	hyperpolarization-activated nonselective cation current
IP_3	inositol 1,4,5-triphosphate
LDS	late developmental stage
NO	nitric oxide
NOS	nitric oxide synthase
PECAM	platelet endothelial cell adhesion molecule
PKA	protein kinase A
VDCC	voltage-dependent L-type Ca^{2+} channel
VEDS	very early developmental stage.

KARGER
Fax + 41 61 306 12 34
E-Mail karger@karger.ch
www.karger.com

1422–6405/99/1654–0153$17.50/0

Accessible online at:
www.karger.com/journals/cto

Dr. Heinrich Sauer
Institute of Neurophysiology
Robert-Koch-Strasse 39
D–50931 Köln (Germany)
Tel. +49 221 478 6976, Fax +49 221 478 6965

to terminally differentiated cells (e.g. ventricular cardiomyocytes, sinus-nodal cells) [Maltsev et al., 1994], the development of endothelial cells giving rise to functional angiogenesis [Vittet et al., 1996; Wartenberg et al., 1998], hematopoiesis [Wiles and Keller, 1991; Kennedy et al., 1997; Nishikawa et al., 1998], development of smooth [Drab et al., 1997] and skeletal [Rohwedel et al., 1994] myocytes as well as of neuronal [Strübing et al., 1995] and glial [Angelov et al., 1998] cells. Besides these mesodermal and ectodermal cell lineages there is also evidence for the formation of a primitive endoderm secreting α-fetoprotein and transferrin [Abe et al., 1996; Sauer et al., 1998]. The ES cell model has recently become even more attractive because of the establishment of human ES cell lines [Thomson et al., 1998; Shamblott et al., 1998]. This may allow the generation of differentiated human cells which eventually may also be used for transplantation purposes.

Another advantage of ES cells is their easy genetic manipulation [Doetschman et al., 1986; Rathjen et al., 1998]. Genes can be knocked out [loss of function, e.g. β_1-integrin knockout ES cells; Fässler and Meyer, 1995; Fässler et al., 1995, 1996], overexpressed or new genes, e.g. reporter gene constructs, may be inserted into the genome of ES cells [Metzger et al., 1996; Wobus et al., 1997; Kolossov et al., 1998]. These models provide powerful tools for investigating the development of functional, physiological as well as pathophysiological aspects of cell proliferation and differentiation.

This review summarizes our work on this model in order to study the relevance of intracellular Ca^{2+} signals, ion channel expression and the establishment of signal transduction cascades for cardiomyogenesis and angiogenesis [Maltsev et al., 1994, 1999; Igelmund et al., 1999; Ji et al., 1999].

The Early Embryoid Body: Spontaneous Ca^{2+} Oscillations in ES Cell-Derived Primitive Endodermal Cells

Murine ES cells have been established from the inner cell mass of murine blastocysts [Evans and Kaufman, 1981]. In the intact embryo the segregation of cells within the inner cell mass involves the formation of the extraembryonic hypoblast, i.e. the primitive endoderm which lines the blastocelic cavity. The remaining cells of the inner cell mass – the epiblast – develop the actual embryo. The most important issues for the developing embryo are (1) to differentiate extraembryonic tissues that enable the embryo to survive in the maternal uterus, and (2) to differentiate embryonic tissues that fulfill its needs for oxygen and nutrients. Consequently, the first organ system to develop in early embryogenesis is the cardiovascular system, which is formed subsequent to the induction of mesodermal cells in the process of gastrulation.

Ca^{2+} oscillations may be the most prominent signals during the very early development. At fertilization periodic Ca^{2+} oscillations release oocytes from meiotic arrest [Kono et al., 1996]. Furthermore, it has recently been shown that oscillations both during exit from meiosis and during mitosis increased the number of inner cell mass cells, which may indicate that Ca^{2+} oscillations have a role in long-term embryonic events [Bos-Mikich et al., 1997]. This led to the suggestion that a Ca^{2+} oscillator is an essential timing element of the early embryonic 'master clock' [Swanson et al., 1997]. In embryoid bodies intracellular Ca^{2+} oscillations were recently reported in cell layers belonging to the primitive endoderm [Sauer et al., 1998]. These cell layers at the periphery of the embryoid body differentiate into the extraembryonic tissues of α-fetoprotein- and transferrin-secreting yolk sac epithelia [Doetschman et al., 1985]. In the embryo the tissue of the visceral yolk sac plays an important role as a protective barrier and as a transporter of selected molecules as well as producing nutritive materials for the developing embryo. The mode by which the embryo transduces its demands for nutrition, growth factors, and hormones to the extraembryonic tissues is currently unknown. It is, however, likely that receptor-mediated signal transduction processes play a key role. These include changes in intracellular Ca^{2+}, which regulate the uptake and transcytotic transport of these substances from the extraembryonic to the embryonic tissues. Single Ca^{2+} spikes occurred already in undifferentiated 3-day-old embryoid bodies and formed regular oscillations from day 4 on (fig. 1). The frequency of Ca^{2+} oscillations in the primitive endoderm-derived cell layers increased up to day 19 of embryoid body culture, whereas the duration of the Ca^{2+} transients decreased. An evaluation of the signal transduction pathways, which direct Ca^{2+} oscillations in visceral yolk sac cell layers, revealed that the Ca^{2+} spikes were mediated by an inositol 1,4,5-triphosphate (IP_3)-dependent pathway with a negative feedback loop via protein kinase C, whereas an involvement of ryanodine receptors was excluded. The endodermal cell layers of embryoid bodies contained numerous endocytotic vesicles which were stained with the oxacarbocyanine fluorescent cell membrane marker DiO (3,3′-dioctadecyloxacarbocyanine perchlorate). Accordingly, it can be assumed that the observed Ca^{2+} oscillations were triggering transcytotic transport processes by mechanisms which may be comparable to those observed in Ca^{2+}-regulated exocytosis in a variety of preparations [Kim et al., 1997; Blank et al., 1998; Jonas et al., 1998; Kwiecien and Hammond, 1998; Tse and Tse, 1998]. We observed that inhibition of Ca^{2+} oscillations by incubation of embryoid bodies in Ca^{2+}-free media, after preincubation with the Ca^{2+} ATPase inhibitor thapsigargin or brefeldin A, which is known to interfere with exocytotic and endocytotic transport processes, prevented endocytosis in embryoid bodies. Under these experimental conditions DiO-stained endocytotic vesicles were absent. These data indicated that Ca^{2+} oscillations trigger the endocytotic vesicle shuttle in embryoid bodies. Since any nutrition from the medium has to pass the extraembryonic epithelial cell layers, i.e. the visceral yolk sac cell layers at the periphery of the embryoid body, a possible role for Ca^{2+} oscillations may be to tune the developing embryoid body's needs for nutrition, vitamins and hormones.

Investigation of Cardiomyogenesis

After formation of early primitive endodermal tissues mediating the supply of nutrients, vitamins and hormones there is a further need to improve the feeding of the embryonic tissue. This is established by switching from diffusion within the embryoid body (presumably via capillaries, see below) to convection within the capillary system, i.e. an active force has to be established in order to provide rhythmic pumping. We have observed cardiac development within embryoid bodies as early as 7 days after formation of the aggregates, which correlates well with the murine embryo where the first beating is seen on day E8.5–E9.5 (i.e. 8.5 to 9.5 days postcoitus) [Fishman and Chien, 1997]. What are the most important functional demands for early cardiomyocytes? Besides the estab-

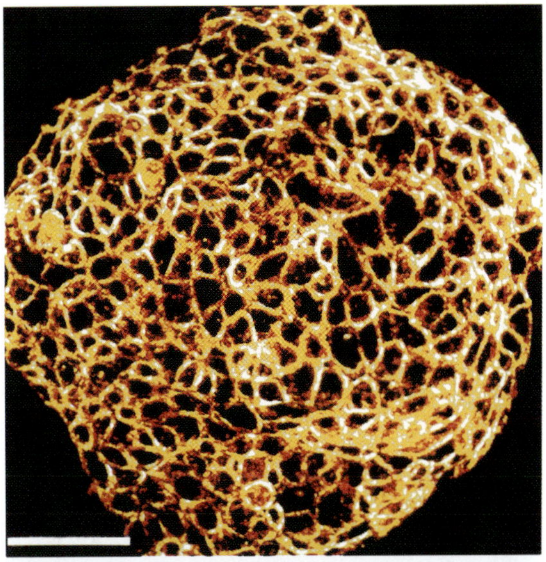
a

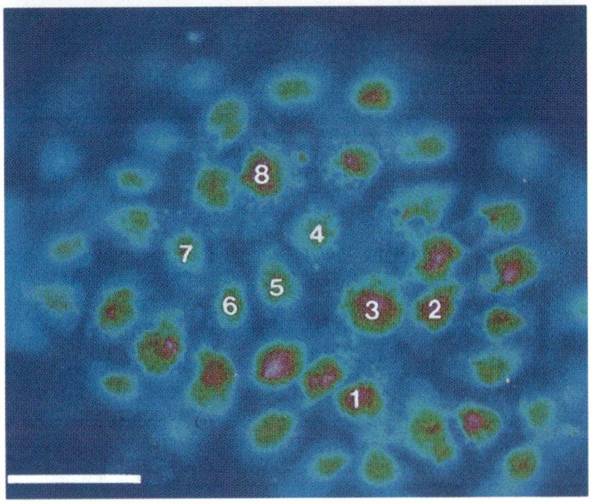

b

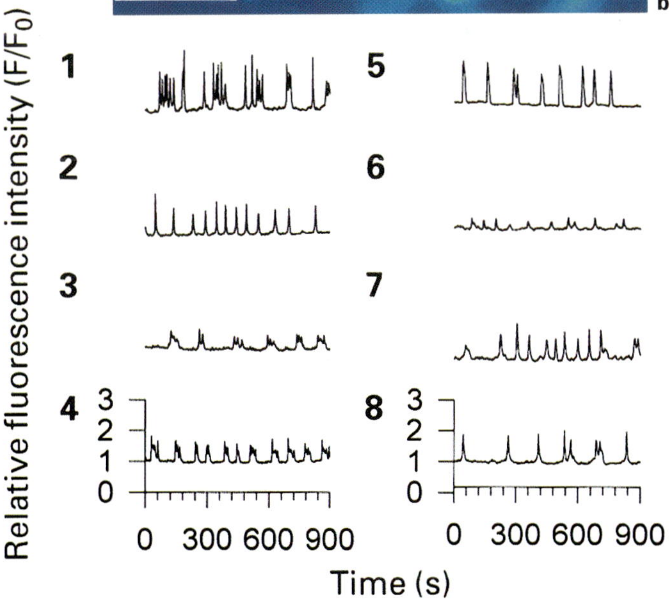

Fig. 1. Spontaneous $[Ca^{2+}]_i$ oscillations in primitive endodermal cells differentiated at the periphery of embryoid bodies. **a** Troma-1 immunostaining of the primitive endodermal cell layer (bar represents 100 μm). **b** Typical tracings of $[Ca^{2+}]_i$ oscillations in individual cells denoted by numbers. Note that the oscillations differ in frequency and amplitude between different cells (bar represents 40 μm) [with permission from Sauer et al., 1998].

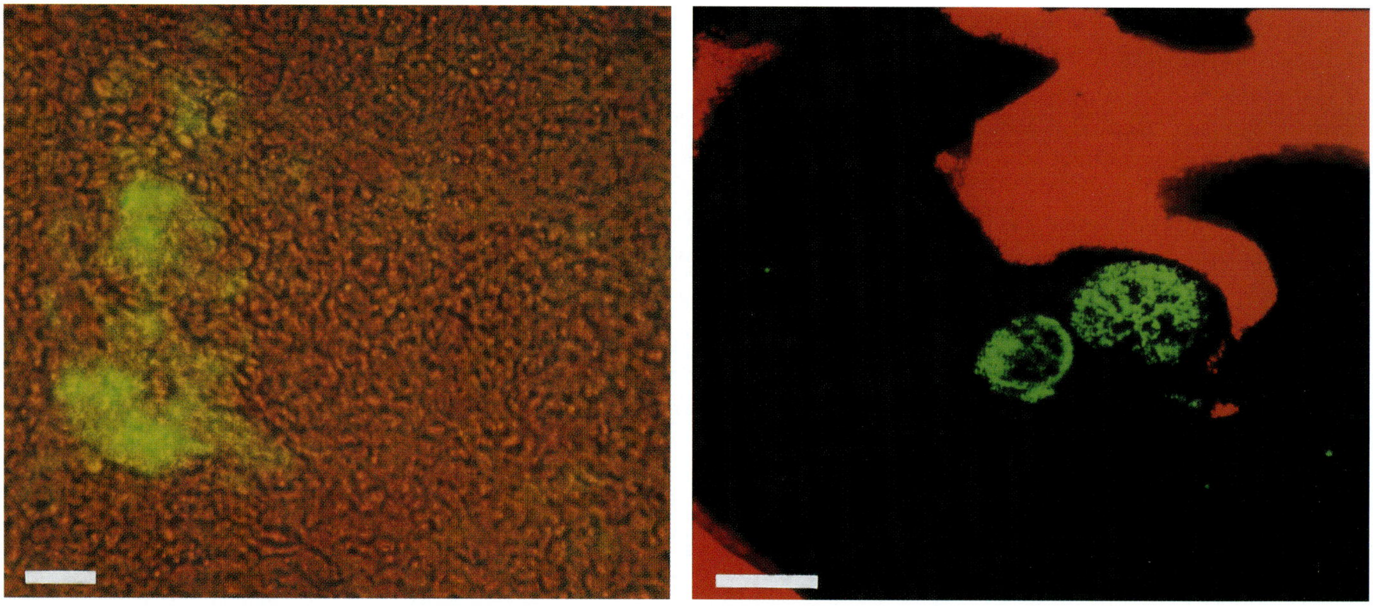

a b

Fig. 2. Expression of GFP under the control of the human cardiac α-actin promotor in cardiomyocytes differentiated from embryonic stem cells (**a**) and the heart of a murine 10.5-day-old embryo (**b**). GFP fluorescence serves to select cardiac precursor cells (cardiac cells prior to the initiation of spontaneous contractions) within embryoid bodies for subsequent electrophysiological and biochemical analysis. Transmission/fluorescence overlay images are shown (bars represent 500 μm).

lishment of the contractile apparatus the cells have to express the cellular machinery required for the generation of the rhythmic contractile activity. Since this aspect is difficult to investigate in the early murine embryonic heart due to the size of the heart prior to E11.5 we have investigated the establishment of the various components involved in excitation-contraction coupling, in particular ion channels and intracellular Ca^{2+} release sites, in ES cell-derived cardiomyocytes. Furthermore, the development-dependent buildup of signalling cascades was also investigated in these cells. Cardiac precursor cells (cardiac cells prior to the initiation of spontaneous contractions) cannot be identified due to the lack of cross striation and of spontaneous beating. Therefore, we established in vivo labeled transgenic ES cell lines.

Generation of Transgenic ES Cells for the Investigation of Cardiac Precursor Cells

The identification of cardiac precursor cells was achieved by generating stably transfected ES cell lines where the in vivo reporter gene GFP (green fluorescent protein) [Prasher et al., 1992; Cody et al., 1993; Chalfie et al., 1994; Heim et al., 1994] was driven by an early cardiac-specific promoter, the human cardiac α-actin promoter [Pari et al., 1991; Sartorelli et al., 1992]. The latter was shown previously to be expressed early during cardiomyogenesis [Metzger et al., 1996]. Upon transfection with electroporation and selection using G418, resistant clones were picked and the differentiation protocol started. Cardiomyocyte-specific GFP expression was observed already starting from very early stages of cardiomyogenesis [Kolossov et al., 1998] (fig. 2a). These in vitro data were further corroborated by establishing – using pronuclear injections – transgenic mice, where the expression of GFP was under control of the same promoter (see fig. 2b). As seen in embryoid bodies the GFP-positive mice started to display cardiac-specific GFP expression already at the earliest stage of the murine heart anlage (E8.5) [Fleischmann et al., 1998].

Table 1. Stage classification and properties of ES cell-derived cardiomyocytes

	Stages				
	0	1	2	3	4
Nomenclature 1	precursor	VEDS	EDS		LDS
Nomenclature 2			early	intermediate	late
Days of differentiation		7 + 1/(2)	7 + (2)/3 to 7 + 4/(5)	7 + (5)/6 to 7 + 8	7 + 9 to 7 + 18
Cell types	uniform	uniform, only one phenotype		diversification of different phenotypes	different phenotypes: ventricular-, atrial-, sinus-nodal-, Purkinje-like cells
Currents	I_{Ca} (10–20 pA at 10 m*M* Ca) Kv channels	I_{Ca}, I_{to}	I_{Ca}, I_{to}, $I_{K,sus}$, I_f(+/–), $I_{K,Ach}$ (slow), $I_{K,ATP}$	specialization of currents	all currents typical for the different phenotypes
Resting potential		depolarized	relatively depolarized, –50 to –70 mV	ND	–80 mV (ventricular) about –45 mV (sinus-nodal)
Membrane potential waves	ND	yes, +++	+/–	–	–
Action potentials	no	small	larger	ND	sinus-nodal-like, ventricular-like, atrial-like, Purkinje-like
Overshoot		no	yes	ND	yes
Contractions	–	+	+	+	+++
Forskolin	–	–	+	++	++
Muscarinic stimulation	–	–	+	+	$I_{K,ACh}$, anti-β-adrenergic
β-Adrenergic stimulation	–	–	–	+/–	+++
NOS–expression			high NOS II, III	ND	low NOS III
cGMP content			high cGMP	ND	low cGMP

Membrane potential waves are characterized by their small amplitude (<30 mV). The hormonal stimulation was evaluated based on its effect onto the amplitude of peak I_{Ca}.
I_{to} = Transient outward rectifying K^+ current; $I_{K,sus}$ = sustained outward rectifying K^+ current;
I_f = hyperpolarization-activated nonselective cation current; $I_{K,ATP}$ = ATP-dependent inward rectifying K^+ current;
Kv = voltage-gated K^+ channel; $I_{K,Ach}$ = acetylcholine-activated inward rectifying K^+ current; ND = not determined.

Functional Expression of the Cellular Components Required for Excitation-Contraction Coupling during Early Cardiomyogenesis

ES cells differentiating to cardiomyocytes within embryoid bodies follow different developmental stages similar as those found in the native embryo. Table 1 summarizes the current knowledge on ion channel expression, pharmacology and modulation in ES cell-derived cardiomyocytes at different developmental stages. The earliest detectable cardiomyocytes (stage '0') using GFP-positive cells (see above) represent a group of cells with relatively homogeneous functional characteristics. These cells were not yet beating but displayed already voltage-dependent L-type Ca^{2+} channel (VDCC) at very low densities (see

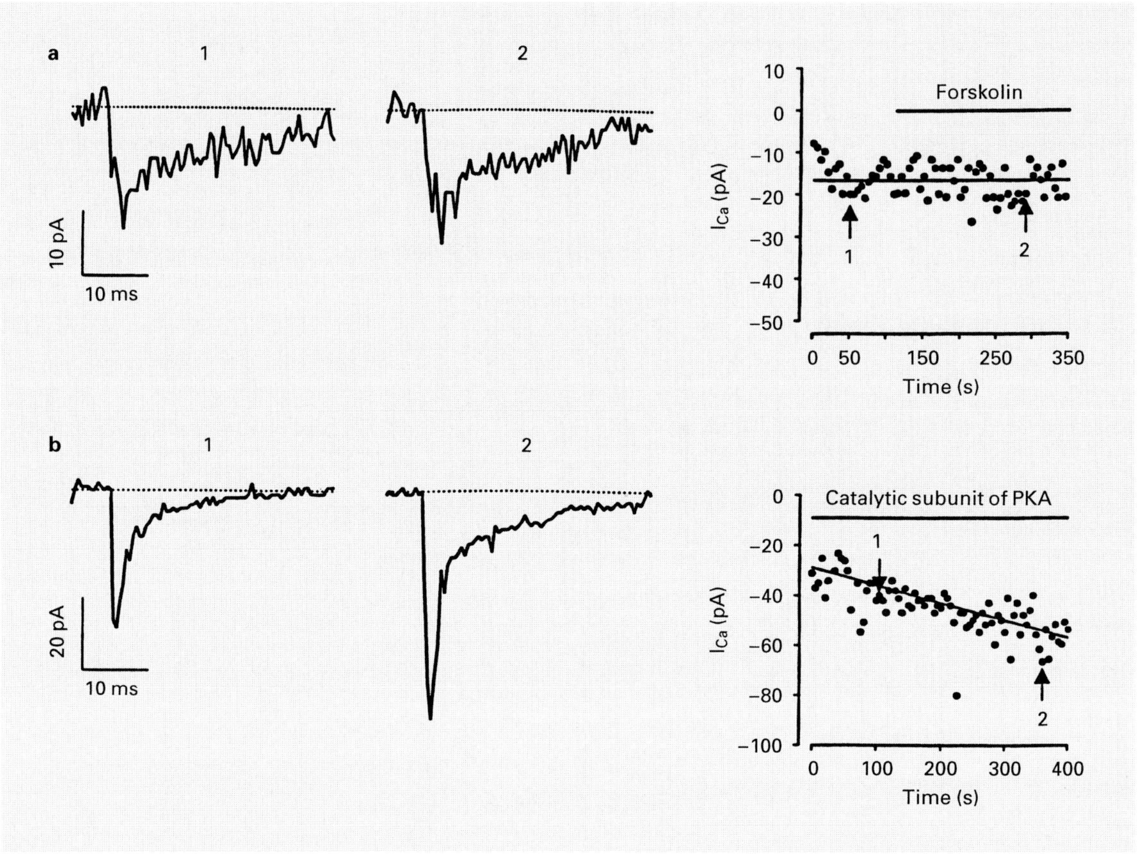

Fig. 3. Pharmacology and regulation of I_{Ca} in 7-day-old, GFP-positive cells: 7 day-old GFP-positive cells (stage '0') were voltage-clamped at −80 mV and 50-ms lasting depolarizing pulses to + 10 mV were applied at a frequency of 0.2 Hz. **a** Superfusion with the adenylyl cyclase activator forskolin (1 μ*M*) did not increase the I_{Ca} amplitude. The mean current level in the time course remains unaltered, confirming that forskolin was without effect. **b** Dialysis with the catalytic subunit of PKA (7 μ*M*) led to an increase of I_{Ca}. This can also be seen from the linear regression displayed in the time course. Every single point in the time course diagrams (right) represents the peak I_{Ca} evoked by a depolarizing voltage step. The numbers on top of the current traces (left diagrams) indicate in the respective time course diagrams (right) when the respective current was recorded [with permission from Kolossov et al., 1998].

below). The next developmental stage, denoted as very early developmental stage (VEDS, stage '1') comprises spontaneously beating cells, still displaying a small set of ion channels, though the most prominent at this stage being VDCC and outwardly rectifying K^+ channels and possibly ATP-sensitive K^+ channels [Gryshchenko and Fleischmann, unpubl. results]. These ion channels generate already primitive action potentials. The early developmental stage (EDS, stage '2') is characterized by an increasing number of functionally expressed ion channels providing stably oscillating action potentials synchronous to contractions. In particular, at this stage distinct subtypes of K^+ channels are functionally expressed (i.e. the transient outward K^+ current, a sustained and a transient outward rectifier K^+ current); moreover, the hyperpolarization-activated nonselective cation current (I_f) is detected in a large percentage of spontaneously beating stage 2 cells [Ji and Fleischmann, unpubl. results]. 7 + 2- as well as 7 + 5-day-old cardiomyocytes are frequently in a transition stage between the VEDS/EDS and the EDS/intermediate stage, respectively. During further development the increasing number of different ion channels causes a diversification of cardiac phenotypes finally leading to the late developmental stage (LDS, stage '4') with the known specialized cardiomyocytes as also found in the neonatal heart. We could identify ventricular-like, atrial-like and sinus-nodal-like as well as Purkinje-like cardiomyocytes [Maltsev et al., 1994; Hescheler et al., 1997]. These cells displayed the regulatory signalling cascades as known in the native counterparts (see table 1) [Maltsev et al., 1994, 1998; Ji et al., 1999].

Since Ca^{2+} homeostasis plays a key role in the regulation of cardiac function [Bers, 1997], we investigated in the GFP-positive cardiac precursor cells (stage '0') as to how Ca^{2+} channels and Ca^{2+} release sites were built up [Kolossov et al., 1998]. Patch clamp and Ca^{2+} imaging experiments proved the functional expression of VDCC starting from day 7. On day 7 I_{Ca} was expressed at a density of 3.8 ± 1.3 pA/pF in the presence of 10 m*M* extracellular Ca^{2+}. Overall, the biophysical and pharmacological properties of VDCC were similar to those of adult cardiomyocytes [Hescheler et al., 1997; Kolossov et al., 1998; Maltsev et al., 1999]. Moreover, even at this early developmental stage I_{Ca} was found to be already stimulated upon cAMP-dependent phosphorylation as reported for terminally differentiated cardiomyocytes [Osterrieder et al., 1982; Hartzell and Fischmeister, 1986; Trautwein and Hescheler, 1990] since intracellular infusion of the catalytic subunit of protein kinase A (PKA) resulted in a 1.7-fold stimulation (fig. 3). Forskolin, a direct activator of the adenylyl cyclase, was without effect. Ca^{2+} imaging experiments proved that IP_3-sensitive intracellular Ca^{2+} stores and Ca^{2+}-ATPases were already present at the ES cell stage. In contrast, functional ryanodine receptors, which were detected by caffeine-induced Ca^{2+} release, appeared 1–2 days later than I_{Ca}, suggesting the lack of Ca^{2+} induced Ca^{2+} release [Fabiato, 1985; Barcenas-Ruiz and Wier, 1987; du Bell and Houser, 1987; Nabauer et al., 1989] in cardiac precursor cells [Kolossov et al., 1998]. The functional expression of the ryanodine receptors coincided with the initiation of the spontaneous contractions. This observation was surprising, since during embryonic cardiomyogenesis electromechanical coupling was reported to be dependent on the transsarcolemmal Ca^{2+} influx [Nayler and Fassold, 1977; Mahony and Jones, 1986; Klitzner and Friedman, 1989; Chin et al., 1990]. Interestingly, murine embryos deficient in the ryanodine 2 receptor die around E11.5, despite the initiation of spontaneous contractions of the heart. This appears to be related to Ca^{2+} overload [Takeshima et al., 1998]. Thus, ryanodine stores during early cardiomyogenesis may be important Ca^{2+} storage sites but not yet required for Ca^{2+} release.

Establishment of Signalling Cascades during Early Cardiomyogenesis

While detailed biophysical as well as in vitro expression studies have provided helpful insight into the functioning of ion channels and the structure-function relationship, investigations on developmental aspects of ion channel function and regulation may give us a better understanding of the complex situation occurring in the terminally differentiated cells. Moreover, this aspect may prove helpful in a better understanding of pathologically transformed cells, since embryonic regulatory mechanisms may be reactivated in these cells [Nakao et al., 1997].

Our group has made substantial progress with the buildup of signalling cascades governing the regulation of ion channels and changes occurring during development. As reported for early murine cardiomyocytes early stage ES cell-derived cardiomyocytes did not express functionally coupled β-adrenoceptors [Slotkin et al., 1994; An et al., 1996], whereas intact muscarinic signalling was present at this stage [Ji et al., 1999]. The most relevant finding was, however, the prominent role of nitric oxide (NO) as a key signalling molecule during early embryonic development [Ji et al., 1999]. In addition, we have identified a switch in the muscarinic regulation during development.

Establishment of Muscarinic Signalling during Cardiomyogenesis

An important observation was the negative chronotropic effect of carbachol (CCh) on spontaneous contractile activity in EDS (stage 2) cells (fig. 4) without hyperpolarization excluding an involvement of the acetylcholine-activated potassium current. These data indicated that the negative chronotropic effect of CCh may be mediated by a depression of I_{Ca}. This was confirmed by voltage clamp experiments, where we confirmed that muscarinic agonists depressed basal VDCC through a pertussis toxin-sensitive G protein and the NO synthase (NOS) III-mediated generation of NO. The NO effect was mediated via activation of the soluble guanylyl cyclase, cGMP, phosphodiesterase-2, decrease of cAMP, PKA activity and the subsequent reduction of I_{Ca}. A direct effect of NO on VDCC [Hu et al., 1997] was excluded since specific inhibitors acting at different sites of the signalling cascade completely blocked the NO effect. Immunocytochemical studies for NOS II and NOS III expression and NADPH-diaphorase staining confirmed our electrophysi-

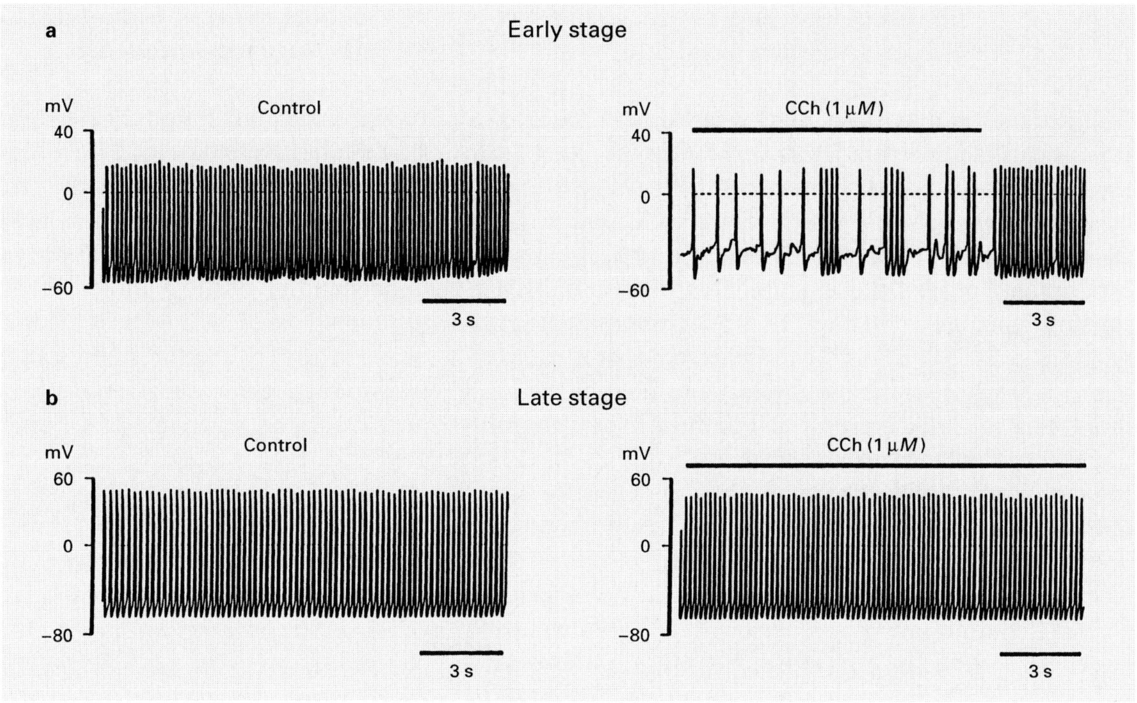

Fig. 4. a Action potentials recorded from an early stage (stage 2), spontaneously contracting cardiomyocyte in the current clamp mode. The cell has a relatively depolarized resting membrane potential and only a small overshoot. In the right panel, CCh was applied exerting a prominent negative chronotropic effect, but no hyperpolarization. The CCh effect can be reversed on washout. **b** Action potentials recorded from a spontaneously contracting, ventricular-like late stage (stage 4) cardiomyocyte. Application of CCh does not change action potential frequency [with permission from Ji et al., 1999].

ological findings [Ji et al., 1999]. Importantly, in the early stage cardiomyocytes high basal activities for adenylyl cyclase but also for the counter regulating systems in the form of phosphodiesterases 2–4 as well as phosphatases were detected [Maltsev et al., 1998; Ji et al., 1999]. Thus, NO is envisioned as a pivotal signalling molecule during cardiomyogenesis [Ji et al., 1999] as reported before for neuronal differentiation [Peunova and Enikolopov, 1995] and for organogenesis in *Drosophila melanogaster* [Kuzin et al., 1996]. Since in the LDS cells there was a direct G_i-mediated depression only of isoprenaline-prestimulated VDCC a switch in the muscarinic regulation of VDCC occurs.

Establishment of β-Adrenergic Signalling during Cardiomyogenesis

Similar to findings at the murine embryonic stage (E11.5–E13.5) we could demonstrate that β-adrenergic signalling was absent at the VEDS/EDS cell stage, despite β-adrenoceptor expression as confirmed by immunocytochemistry [Bloch, unpubl. results]. However, even at this stage VDCC can be stimulated by the catalytic subunit of PKA to the same degree as in LDS cells proving that the phosphorylation site of VDCC or a closely related protein are already functioning [Maltsev et al., 1999]. The defect at this early stage does not appear to be solely related to a possible defect in β-adrenoceptor-G_s coupling, since application of 8-bromo-cAMP or forskolin does also not or only weakly stimulate VDCC [An et al., 1996; Maltsev et al., 1999]. Then, a fast transition to the complete establishment of the β-adrenergic signalling cascade is observed. It is still unclear as to whether the defective signalling is related to a deficiency of the PKA holoenzyme [An et al., 1996] or rather the counterbalancing actions of high intrinsic phosphodiesterase and phosphatase activity [Maltsev et al., 1999]. Interestingly, the direct G_s-mediated stimulation of VDCC [Yatani and Brown, 1989] also only functions in LDS cells.

Functional Aspects of Angiogenesis in Embryoid Bodies

The distribution of oxygen, growth factors and nutrients is of predominant importance for a growing tissue. In unvascularized tissues, e.g. avascular tumors, gradients in oxygen and nutrition as well as accumulation of toxic catabolic end products in deeper cell layers result in necrosis and tissue growth arrest [Freyer and Sutherland, 1986]. Hence, the cardiovascular system is the first organ system to develop in embryos and is already formed during gastrulation when the inner cell mass invaginates through the primitive streak and mesodermal cells are induced during this process [Risau, 1995]. Progenitors of blood cells and endothelial cells differentiate from the mesoderm at EDS. Endothelial cells arise from between stage E7.0 and E8.5 of murine development and differentiate to angioblasts which are located within the paraxial and lateral plate mesoderm and within the yolk sac extraembryonic mesoderm where they surround blood cell islands [Risau, 1997]. The formation of these blood vessels from differentiating angioblasts is called vasculogenesis, whereas during the subsequent process of angiogenesis blood islands fuse, form lumina and sprout new capillaries from the preexisting primitive vascular plexus.

All essential steps of hematopoiesis, vasculogenesis and angiogenesis have been observed in embryoid bodies derived from murine ES cells [Wiles and Keller, 1991; Doetschman et al., 1993; Vittet et al., 1996; Kennedy et al., 1997; Wartenberg et al., 1998]. Endothelial cell differentiation in embryoid bodies occurs spontaneously without the addition of growth factors such as basic or acidic fibroblast growth factor or vascular endothelial growth factor. This indicates that these growth factors, which have been demonstrated to be essential for endothelial cell differentiation, are produced by the embryoid body tissue. The endothelial cells differentiated in embryoid bodies have been characterized by expression studies of various endothelial cell-specific molecules [Vittet et al., 1996]. These markers included growth factor receptors, i.e. Flk-1, tie-1, and tie-2, adhesive molecules such as vascular endothelial cadherin, platelet endothelial cell adhesion molecule (PECAM-1), as well as other endothelial cell-specific antigens such as MECA-32 and MEC.14.7.

In a recent study of our group we introduced a new spinner flask culture technique to optimize the differentiation of endothelial cells in embryoid bodies [Wartenberg et al., 1998]. This culture technique appeared to be superior to traditional culture techniques since it resulted in greater differentiation efficiency, and less time was required to achieve differentiation. Using PECAM-1 as a marker for endothelial cell differentiation we observed by immunohistochemical methods that single PECAM-1-positive cells were present in embryoid bodies on day 4 of cell culture, whereas cell clusters which presumably represented angioblasts were present on day 5. From the angioblastic PECAM-1-positive cell clusters sprouting endothelial cells were observed on day 6. Network-like capillary structures were formed on day 7–8 in 100% of the investigated embryoid bodies. After day 8 no further increase of the capillary area in embryoid bodies was observed. This implicates that angiogenesis was finished at this developmental stage and endothelial cells entered cell quiescence which is the normal situation in adult tissues where endothelial outgrowth from pre-existing vessels is a relatively rare event, occurring only during wound healing and menstruation. In parallel to the termination of angiogenesis, embryoid body growth declined to plateau levels, indicating that supply with oxygen and nutrients in the embryoid bodies had reached homeostatic levels. In embryoid bodies which contain capillary structures, development of central necrosis as observed in equal-sized multicellular tumor spheroids is absent [Wartenberg and Acker, 1995; Gassmann et al., 1996]. This points towards the notion that the capillaries in embryoid bodies form functional vessels which improve diffusion within the tissue and detoxify the tissue from catabolic end products (fig. 5).

To evaluate the diffusion properties of the tissue we incubated embryoid bodies with the cell permeant fluorescent anthracycline doxorubicin and evaluated the diffusion constant for this compound in avascular 3-day-old embryoid bodies, vascularized 8-day-old embryoid bodies and embryoid bodies treated with agents which inhibited angiogenesis including tamoxifen, suramin and heparin/tetrahydrocortisol. Our data indicate that in vascularized embryoid bodies doxorubicin diffusion is about 10-fold increased as compared to avascular embryoid bodies, which indicates that in embryoid bodies functional vessels are formed. These data were corroborated by experiments where avascular and vascularized embryoid bodies were incubated by a cell-impermeant 10-kD dextran coupled to the fluorophor BCECF [2′,7′-bis-(2-carboxyethyl)-5-(and-6)-carboxyfluorescein]. In avascular embryoid bodies the dextran was excluded whereas in vascularized embryoid bodies vascular structures filled with fluorescent dextran were obvious. The development of a luminized tube is a key event during vasculogenesis and angio-

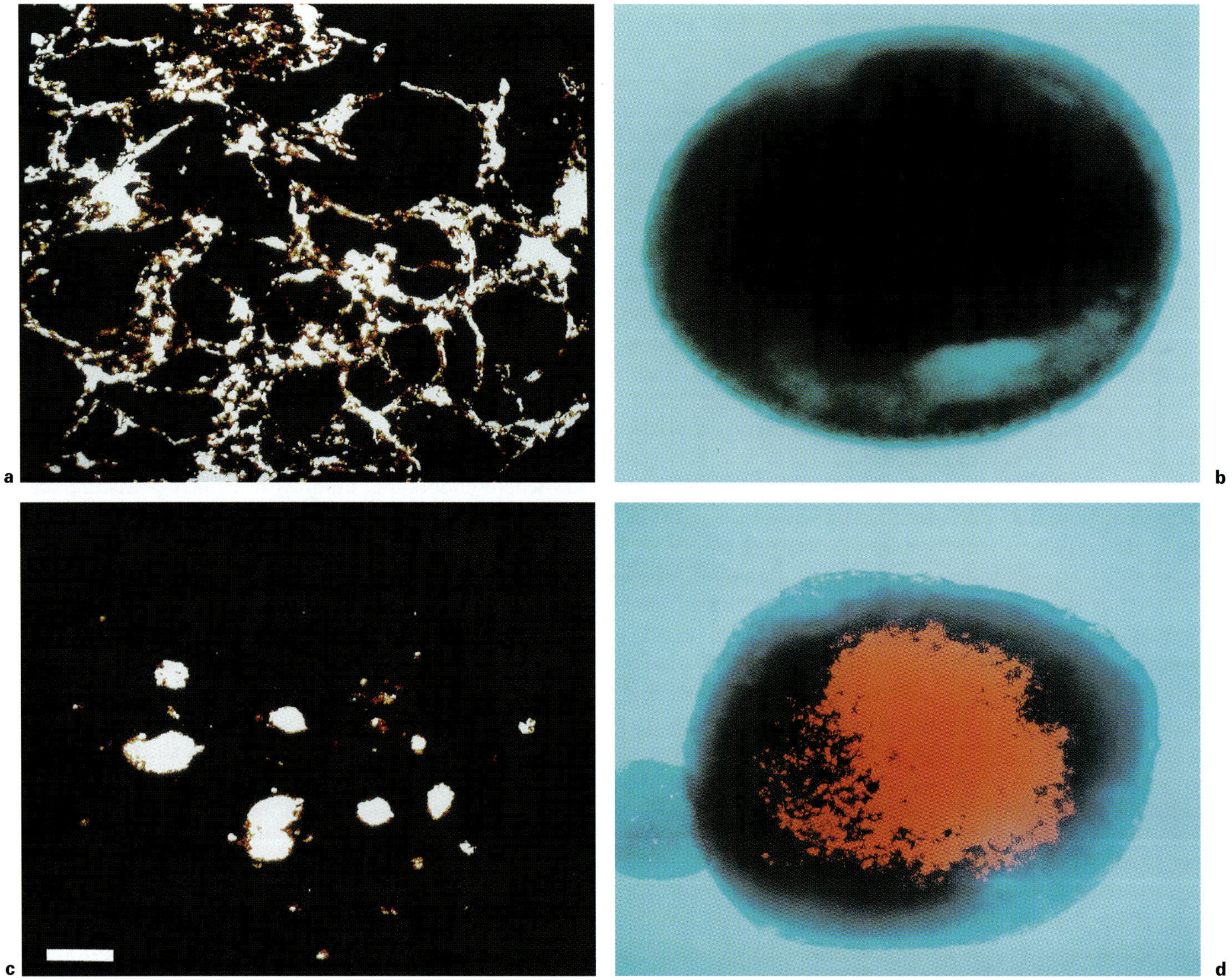

Fig. 5. Inhibition of angiogenesis and necrosis development in embryoid bodies by tamoxifen. **a**, **c** Capillary structures immunostained with an antibody directed against PECAM-1. **a** Untreated control on day 8 of development. **c** Embryoid body on day 8 of development after treatment for 4 days with tamoxifen. **b**, **d** Necrosis staining of control (**b**) and tamoxifen (**d**)-treated embryoid bodies using lucifer yellow VS (4-amino-N-[3-(vinylsulfonyl)-phenyl]naphthalimide-3,6-disulfonate). Transmission/fluorescence overlay images (blue color = transmission, red color = lucifer yellow staining). Note that necrosis is absent in vascularized embryoid bodies, whereas pronounced central necrosis occurs after treatment with the antiangiogenic agent tamoxifen. (bar represents 100 µm) [with permission from Wartenberg et al., 1998].

genesis. Lumen formation is a multifactorial process which requires cell-cell contacts mediated by E-selectin, by cadherins, PECAM-1 and integrins. Since mechanical and chemical signals are transduced through the extracellular matrix and the surrounding tissue cytoarchitecture, investigation of vasculogenesis and angiogenesis needs complex in vivo-like conditions which are absent in conventional endothelial cell cultures but are ideally provided by the tissue of embryoid bodies.

References

Abe, K., H. Niwa, K. Iwase, M. Takiguchi, M. Mori, S.I. Abe, K.I. Yamamura (1996) Endoderm-specific gene expression in embryonic stem cells differentiated to embryoid bodies. Exp Cell Res *229:* 27–34.

An, R.H., M.P. Davies, P.A. Doevendans, S.W. Kubalak, R. Bangalore, K.R. Chien, R.S. Kass (1996) Developmental changes in beta-adrenergic modulation of L-type Ca^{2+} channels in embryonic mouse heart. Circ Res *78:* 371–378.

Angelov, D.N., S. Arnhold, C. Andressen, H. Grabsch, M. Puschmann, J. Hescheler, K. Addicks (1998) Temporospatial relationship between macroglia and microglia during in vitro differentiation of murine stem cells. Dev Neurosci *20:* 42–51.

Barcenas-Ruiz, L., W.G. Wier (1987) Voltage dependence of intracellular $[Ca^{2+}]_i$ transients in guinea pig ventricular myocytes. Circ Res *61:* 148–154.

Bers, D.M. (1997) Ca transport during contraction and relaxation in mammalian ventricular muscle. Basic Res Cardiol *92 (suppl 1):* 1–10.

Blank, P.S., M.S. Cho, S.S. Vogel, D. Kaplan, A. Kang, J. Malley, J. Zimmerberg (1998) Submaximal responses in calcium-triggered exocytosis are explained by differences in the calcium sensitivity of individual secretory vesicles. J Gen Physiol *112:* 559–567.

Bos-Mikich, A., D.G. Whittingham, K.T. Jones (1997) Meiotic and mitotic Ca^{2+} oscillations affect cell composition in resulting blastocysts. Dev Biol *182:* 172–179.

Chalfie, M., Y. Tu, G. Euskirchen, W.W. Ward, D.C. Prasher (1994) Green fluorescent protein as a marker for gene expression. Science *263:* 802–805.

Chin, T.K., W.F. Friedman, T.S. Klitzner (1990) Developmental changes in cardiac myocyte calcium regulation. Circ Res *67:* 574–579.

Cody, C.W., D.C. Prasher, W.M. Westler, F.G. Prendergast, W.W. Ward (1993) Chemical structure of the hexapeptide chromophore of the Aequorea green-fluorescent protein. Biochemistry *32:* 1212–1218.

Doetschman, T., A. Gossler, E. Serfling, W. Schaffner, K. Marcu, L. Stanton, R. Kemler (1986) Introduction of genes into mouse embryonic stem cells. Prog Clin Biol Res *217A:* 47–50.

Doetschman, T., M. Shull, A. Kier, J.D. Coffin (1993) Embryonic stem cell model systems for vascular morphogenesis and cardiac disorders. Hypertension *22:* 618–629.

Doetschman, T.C., H. Eistetter, M. Katz, W. Schmidt, R. Kemler (1985) The in vitro development of blastocyst-derived embryonic stem cell lines: Formation of visceral yolk sac, blood islands and myocardium. J Embryol Exp Morphol *87:* 27–45.

Drab, M., H. Haller, R. Bychkov, B. Erdmann, C. Lindschau, H. Haase, I. Morano, F.C. Luft, A.M. Wobus (1997) From totipotent embryonic stem cells to spontaneously contracting smooth muscle cells: A retinoic acid and db-cAMP in vitro differentiation model. FASEB J *11:* 905–915.

du Bell, W.H., S.R. Houser (1987) A comparison of cytosolic free Ca^{2+} in resting feline and rat ventricular myocytes. Cell Calcium *8:* 259–268.

Evans, M.J., M.H. Kaufman (1981) Establishment in culture of pluripotential cells from mouse embryos. Nature *292:* 154–156.

Fabiato, A. (1985) Effects of ryanodine in skinned cardiac cells. Fed Proc *44:* 2970–2976.

Fässler, R., M. Meyer (1995) Consequences of lack of beta 1 integrin gene expression in mice. Genes Dev *9:* 1896–1908.

Fässler, R., M. Pfaff, J. Murphy, A.A. Noegel, S. Johansson, R. Timpl, R. Albrecht (1995) Lack of beta 1 integrin gene in embryonic stem cells affects morphology, adhesion, and migration but not integration into the inner cell mass of blastocysts. J Cell Biol *128:* 979–988.

Fässler, R., J. Rohwedel, V. Maltsev, W. Bloch, S. Lentini, K. Guan, D. Gullberg, J. Hescheler, K. Addicks, A.M. Wobus (1996) Differentiation and integrity of cardiac muscle cells are impaired in the absence of beta 1 integrin. J Cell Sci *109:* 2989–2999.

Fishman, M.C., K.R. Chien (1997) Fashioning the vertebrate heart: Earliest embryonic decisions. Development *124:* 2099–2117.

Fleischmann, M., W. Bloch, E. Kolossov, C. Andressen, M. Muller, G. Brem, J. Hescheler, K. Addicks, B.K. Fleischmann (1998) Cardiac specific expression of the green fluorescent protein during early murine embryonic development. FEBS Lett *440:* 370–376.

Freyer, J.P., R.M. Sutherland (1986) Regulation of growth saturation and development of necrosis in EMT6/Ro multicellular spheroids by the glucose and oxygen supply. Cancer Res *46:* 3504–3512.

Gassmann, M., J. Fandrey, S. Bichet, M. Wartenberg, H.H. Marti, C. Bauer, R.H. Wenger, H. Acker (1996) Oxygen supply and oxygen-dependent gene expression in differentiating embryonic stem cells. Proc Natl Acad Sci USA *93:* 2867–2872.

Gossler, A., T. Doetschman, R. Korn, E. Serfling, R. Kemler (1986) Transgenesis by means of blastocyst-derived embryonic stem cell lines. Proc Natl Acad Sci USA *83:* 9065–9069.

Hartzell, H.C., R. Fischmeister (1986) Opposite effects of cyclic GMP and cyclic AMP on Ca^{2+} current in single heart cells. Nature *323:* 273–275.

Heim, R., D.C. Prasher, R.Y. Tsien (1994) Wavelength mutations and posttranslational autoxidation of green fluorescent protein. Proc Natl Acad Sci USA *91:* 12501–12504.

Hescheler, J., B.K. Fleischmann, S. Lentini, V.A. Maltsev, J. Rohwedel, A.M. Wobus, K. Addicks (1997) Embryonic stem cells: A model to study structural and functional properties in cardiomyogenesis. Cardiovasc Res *36:* 149–162.

Hu, H., N. Chiamvimonvat, T. Yamagishi, E. Marban (1997) Direct inhibition of expressed cardiac L-type Ca^{2+} channels by S-nitrosothiol nitric oxide donors. Circ Res *81:* 742–752.

Igelmund, P., B.K. Fleischmann, I.R. Fischer, J. Soest, O. Gryshchenko, M.M. Bohm-Pinger, H. Sauer, Q. Liu, J. Hescheler (1999) Action potential propagation failures in long-term recordings from embryonic stem-cell derived cardiomyocytes in tissue culture. Pflügers Arch *437:* 669–679.

Ji, G.J., B.K. Fleischmann, W. Bloch, M. Feelisch, C. Andressen, K. Addicks, J. Hescheler (1999) Regulation of the L-type Ca^{2+}-channel during cardiomyogenesis: Switch from NO to adenylyl-cyclase-mediated inhibition. FASEB J *13:* 313–324.

Jonas, J.C., P. Gilon, J.C. Henquin (1998) Temporal and quantitative correlations between insulin secretion and stably elevated or oscillatory cytoplasmic Ca^{2+} in mouse pancreatic beta-cells. Diabetes *47:* 1266–1273.

Kennedy, M., M. Firpo, K. Choi, C. Wall, S. Robertson, N. Kabrun, G. Keller (1997) A common precursor for primitive erythropoiesis and definitive haematopoiesis. Nature *386:* 488–493.

Kim, T.D., G.T. Eddlestone, S.F. Mahmoud, J. Kuchtey, C. Fewtrell (1997) Correlating Ca^{2+} responses and secretion in individual RBL-2H3 mucosal mast cells. J Biol Chem *272:* 31225–31229.

Klitzner, T.S., W.F. Friedman (1989) A diminished role for the sarcoplasmic reticulum in newborn myocardial contraction: Effects of ryanodine. Pediatr Res *26:* 98–101.

Kolossov E., B.K. Fleischmann, Q. Liu, W. Bloch, S. Viatchenko-Karpinski, O. Manzke, G.J. Ji, H. Bohlen, K. Addicks, J. Hescheler (1998) Functional characteristics of ES cell-derived cardiac precursor cells identified by tissue specific expression of the green fluorescent protein. J Cell Biol *143:* 2045–2056.

Kono, T., K.T. Jones, A. Bos-Mikich, D.G. Whittingham, J. Carroll (1996) A cell cycle-associated change in Ca^{2+} releasing activity leads to the generation of Ca^{2+} transients in mouse embryos during the first mitotic division. J Cell Biol *132:* 915–923.

Kuzin, B., I. Roberts, N. Peunova, G. Enikolopov (1996) Nitric oxide regulates cell proliferation during drosophila development. Cell *87:* 639–649.

Kwiecien, R., C. Hammond (1998) Differential management of Ca^{2+} oscillations by anterior pituitary cells: A comparative overview. Neuroendocrinology *68:* 135–151.

Mahony, L., L.R. Jones (1986) Developmental changes in cardiac sarcoplasmic reticulum in sheep. J Biol Chem *261:* 15257–15265.

Maltsev, V.A., G.J. Ji, A.M. Wobus, B.K. Fleischmann, J. Hescheler (1999) Establishment of β-adrenergic modulation of L-type calcium current in the early stages of cardiomyocyte development. Circ Res *84:* 136–145.

Maltsev, V.A., A.M. Wobus, J. Rohwedel, M. Bader, J. Hescheler (1994) Cardiomyocytes differentiated in vitro from embryonic stem cells developmentally express cardiac-specific genes and ionic currents. Circ Res *75:* 233–244.

Metzger, J.M., W.I. Lin, L.C. Samuelson (1996) Vital staining of cardiac myocytes during embryonic stem cell cardiogenesis in vitro. Circ Res *78:* 547–552.

Nabauer, M., G. Callewaert, L. Cleemann, M. Morad (1989) Regulation of calcium release is gated by calcium current, not gating charge, in cardiac myocytes. Science *244:* 800–803.

Nakao, K., W. Minobe, R. Roden, M.R. Bristow, L.A. Leinwand (1997) Myosin heavy chain gene expression in human heart failure. J Clin Invest *100:* 2362–2370.

Nayler, W.G., E. Fassold (1977) Calcium accumulating and ATPase activity of cardiac sarcoplasmic reticulum before and after birth. Cardiovasc Res *11:* 231–237.

Nishikawa, S.I., S. Nishikawa, M. Hirashima, N. Matsuyoshi, H. Kodama (1998) Progressive lineage analysis by cell sorting and culture identifies FLK1+VE-cadherin+ cells at a diverging point of endothelial and hemopoietic lineages. Development *125:* 1747–1757.

Osterrieder, W., G. Brum, J. Hescheler, W. Trautwein, V. Flockerzi, F. Hofmann (1982) Injection of subunits of cyclic AMP-dependent protein kinase into cardiac myocytes modulates Ca^{2+} current. Nature *298:* 576–578.

Pari, G., K. Jardine, M.W. McBurney (1991) Multiple CArG boxes in the human cardiac actin gene promoter required for expression in embryonic cardiac muscle cells developing in vitro from embryonal carcinoma cells. Mol Cell Biol *11:* 4796–4803.

Peunova, N., G. Enikolopov (1995) Nitric oxide triggers a switch to growth arrest during differentiation of neuronal cells. Nature *375:* 68–73.

Prasher, D.C., V.K. Eckenrode, W.W. Ward, F.G. Prendergast, M.J. Cormier (1992) Primary structure of the *Aequorea victoria* green-fluorescent protein. Gene *111:* 229–233.

Rathjen, P.D., J. Lake, L.M. Whyatt, M.D. Bettess, J. Rathjen (1998) Properties and uses of embryonic stem cells: Prospects for application to human biology and gene therapy. Reprod Fertil Dev *10:* 31–47.

Risau, W. (1995) Differentiation of endothelium. FASEB J *9:* 926–933.

Risau, W. (1997) Mechanisms of angiogenesis. Nature *386:* 671–674.

Rohwedel, J., V. Maltsev, E. Bober, H.H. Arnold, J. Hescheler, A.M. Wobus (1994) Muscle cell differentiation of embryonic stem cells reflects myogenesis in vivo: Developmentally regulated expression of myogenic determination genes and functional expression of ionic currents. Dev Biol *164:* 87–101.

Sartorelli, V., N.A. Hong, N.H. Bishopric, L. Kedes (1992) Myocardial activation of the human cardiac alpha-actin promoter by helix-loop-helix proteins. Proc Natl Acad Sci USA *89:* 4047–4051.

Sauer, H., C. Hofmann, M. Wartenberg, A.M. Wobus, J. Hescheler (1998) Spontaneous calcium oscillations in embryonic stem cell-derived primitive endodermal cells. Exp Cell Res *238:* 13–22.

Shamblott, M.J., J. Axelmann, S. Wang, E.M. Bugg, J.W. Littlefield, P.J. Donovan, P.D. Blumenthal, G.R. Huggins, J.D. Gearhart (1998) Derivation of pluripotent stem cells from cultured human primordial germ cells. Proc Natl Acad Sci USA *95:* 13726–13731.

Slotkin, T.A., C. Lau, F.J. Seidler (1994) Beta-adrenergic receptor overexpression in the fetal rat: Distribution, receptor subtypes, and coupling to adenylate cyclase activity via G-proteins. Toxicol Appl Pharmacol *129:* 223–234.

Strübing, C., G. Ahnert Hilger, J. Shan, B. Wiedenmann, J. Hescheler, A.M. Wobus (1995) Differentiation of pluripotent embryonic stem cells into the neuronal lineage in vitro gives rise to mature inhibitory and excitatory neurons. Mech Dev *53:* 275–287.

Swanson, C.A., A.P. Arkin, J. Ross (1997) An endogenous calcium oscillator may control early embryonic division. Proc Natl Acad Sci USA *94:* 1194–1199.

Takeshima, H., S. Komazaki, K. Hirose, M. Nishi, T. Noda, M. Iino (1998) Embryonic lethality and abnormal cardiac myocytes in mice lacking ryanodine receptor type 2. EMBO J *17:* 3309–3316.

Thomson, J.A., J. Itskovitz-Eldo, S.S. Shapiro, M.A. Waknitz, J.J. Swiergiel, V.S. Marshall, J.M. Jones (1998) Embryonic stem cell lines derived from human blastocysts. Science *282:* 1145–1147.

Trautwein, W., J. Hescheler (1990) Regulation of cardiac L-type calcium current by phosphorylation and G proteins. Annu Rev Physiol *52:* 257–274.

Tse, A., F.W. Tse (1998) Alpha-adrenergic stimulation of cytosolic Ca^{2+} oscillations and exocytosis in identified rat corticotrophs. J Physiol (Lond) *512:* 385–393.

Vittet, D., M.H. Prandini, R. Berthier, A. Schweitzer, H. Martin-Sisteron, G. Uzan, E. Dejana (1996) Embryonic stem cells differentiate in vitro to endothelial cells through successive maturation steps. Blood *88:* 3424–3431.

Wartenberg, M., H. Acker (1995) Quantitative recording of vitality patterns in living multicellular spheroids by confocal microscopy. Micron *26:* 395–404.

Wartenberg, M., J. Gunther, J. Hescheler, H. Sauer (1998) The embryoid body as a novel in vitro assay system for antiangiogenic agents. Lab Invest *78:* 1301–1314.

Wiles, M.V., G. Keller (1991) Multiple hematopoietic lineages develop from embryonic stem (ES) cells in culture. Development *111:* 259–267.

Wobus, A.M., G. Kaomei, J. Shan, M.C. Wellner, J. Rohwedel, G.J. Ji, B. Fleischmann, H.A. Katus, J. Hescheler, W.M. Franz (1997) Retinoic acid accelerates embryonic stem cell-derived cardiac differentiation and enhances development of ventricular cardiomyocytes. J Mol Cell Cardiol *29:* 1525–1539.

Wobus, A.M., G. Wallukat, J. Hescheler (1991) Pluripotent mouse embryonic stem cells are able to differentiate into cardiomyocytes expressing chronotropic responses to adrenergic and cholinergic agents and Ca^{2+} channel blockers. Differentiation. *48:* 173–182.

Yatani, A., A.M. Brown (1989) Rapid beta-adrenergic modulation of cardiac calcium channel currents by a fast G protein pathway. Science *245:* 71–74.

Cells
Tissues
Organs

Cells Tissues Organs 1999;165:165–172

Accepted after revision: March 29, 1999

An in vitro Pathway from Embryonic Stem Cells to Neurons and Glia

David I. Gottlieb   James E. Huettner

Department of Anatomy and Neurobiology and Department of Cell Biology and Physiology, Washington University School of Medicine, St. Louis, Mo., USA

Key Words
Embryonic stem cells · Neurons · Glia

Abstract
Mouse embryonic stem (ES) cells can be induced to differentiate into neurons and glia in vitro. Induction protocols are straightforward and involve culture in the presence of retinoic acid. They result in an efficient conversion of undifferentiated ES cells to neural cells. Mature neurons produced have the key physiological, morphological and molecular properties of primary cultured neurons derived from the central nervous system. Most significantly, they form functional chemical synapses that utilize either glutamate, GABA or glycine as neurotransmitters. ES cell-derived glial cells also correspond well with their normal counterparts. During induction, ES cells undergo a series of developmental steps that resemble key stages in the early mouse embryo. This supports the hypothesis that the in vitro pathway is a valid model of the normal developmental pathway leading to neurons and glia. The in vitro system combines three experimental strengths. It is suitable for genetic manipulation, affords large numbers of cells and allows precise manipulation of the culture environment. It is thus suitable for a wide variety of mechanistic studies in the areas of neural development and cell biology.

Introduction

Mouse embryonic stem (ES) cells strongly resemble cells of the inner cell mass and primitive ectoderm, which are transient structures found in the embryonic day 4–6 (E4–6) embryo. The primitive ectoderm, which is the source of all cells of the body, gives rise to the neural plate by E8. The neural plate is quickly transformed into the neural tube. At early stages, the neural tube consists predominantly of rapidly dividing stem cells. Stem cells later exit the cell cycle to give rise to the neurons and glia of the nervous system. ES cells have the inherent capacity to differentiate into all cell types of the nervous system as demonstrated by formation of chimeric mice from embryos in which ES cells are transplanted into the inner cell mass. At

Abbreviations used in this paper

AMPA	amino-3-hydroxy-5-methylisoxazole-4-propionic acid
bFGF	basic fibroblast growth factor
E	embryonic
EB	embryoid body
ES cells	embryonic stem cells
GABA	γ-aminobutyric acid
GAD	glutamic acid decarboxylase
LIF	leukemia inhibitory factor
NMDA	N-methyl-*D*-aspartate
RA	retinoic acid

KARGER
Fax +41 61 306 12 34
E-Mail karger@karger.ch
www.karger.com

1422–6405/99/1654–0165$17.50/0

Accessible online at:
www.karger.com/journals/cto

Prof. David Gottlieb
Department of Anatomy and Neurobiology
Campus Box 8108, Washington University School of Medicine
660 S. Euclid Avenue, St. Louis, MO 63110 (USA)
Tel. +1 314 362 2758, Fax +1 314 362 3446, E-Mail gottlied@thalamus.wustl.edu

first glance, it would appear unlikely that the developmental lineage extending from ES cells to mature neurons and glia could be recapitulated in cell culture because the processes which give rise to the vertebrate nervous system, while not understood in detail, appear to be highly complex. Contrary to this expectation, it has proved possible to reconstitute, at least partially, this pathway in vitro starting with ES cells and ending with differentiated neurons and glia. The result is an experimental system combining three strengths. Genetic manipulation is available because the ES cell is the leading model for targeted modification of the genome. Since ES cells are a large-scale source of cells, biochemical investigation is feasible. Finally, tissue culture makes it possible to alter the environment of cells in precise ways. Thus studies with the system have the potential for making important contributions to understanding neuronal and glial development and function. This review covers the literature in which ES cells have been induced to differentiate along a neural lineage in cell culture.

ES cells are maintained in vitro as totipotent stem cells by culture in the presence of the cytokine leukemia inhibitory factor (LIF). Most studies of in vitro differentiation of ES cells begin by culturing ES cells as small aggregates called embryoid bodies (EBs). EBs are formed by withdrawing LIF and culturing cells in dishes with a non-adhesive substratum. Standard EB cultures are composed of a wide variety of cell types, including a small proportion with a neuronal appearance [see for example Martin, 1981]. Three papers published in 1995 demonstrated that more efficient in vitro differentiation of ES cells into neural cells is possible. All utilized retinoic acid (RA) but differ in the details of culture method. In our laboratories [Bain et al., 1995], ES cells were cultured as EBs in the absence of LIF for 4 days. Culture was then continued for 4 additional days in the presence of RA. Next, aggregates were dissociated with trypsin and plated as monolayer cultures on an adhesive substrate. Within several days, large numbers of neuron-like cells appeared. The phenotype of these cells (reviewed below) justifies designating them as neurons. About 40% of the cells in the cultures were neurons. These grew on top of a layer of flat cells adhering to the substrate. Many of the flat cells were glia, but the exact percentage has not been determined. A study by Strübing et al. [1995] used a similar approach. EBs were formed by the 'hanging drop' culture method. These EBs were exposed to RA for the first 2 days of culture and then cultured in suspension for an additional 2 days in the absence of RA. Next, the EBs were cultured on an adhesive substrate; neurons appeared about 2 days after plating on the adhesive substrate. RA had a dramatic effect. All EBs exposed to RA gave rise to neurons whereas only 15% of control EBs not exposed to RA had neurons. In the study of Fraichard et al. [1995], EBs were formed in mass culture and exposed to RA for the first 2 days. They were then plated onto an adhesive substrate in the absence of RA. After 4–5 days, neuronal cells appeared. Thus three related protocols involving culture of EBs in the presence of RA result in extensive neuronal differentiation. Another approach to obtaining neurons from ES cells has been described [Okabe et al., 1996]. In this case, ES cells were cultured as EBs for 4 days in standard medium in the absence of added RA. Next the EBs were transferred to an adhesive substrate, to which they attach, and the standard medium replaced by a serum-free medium. Over the next several days, many cells die. Many of the surviving cells are tightly packed epithelial cells that are nestin positive. In the presence of basic fibroblast growth factor (bFGF) these cells proliferate and go on to differentiate into neurons and glia.

Physiological Properties of ES Cell-Derived Neurons

The neurons derived from in vitro differentiation of ES cells express voltage- and ligand-gated ion channels. In mature cultures, nearly all of the cells with neuronal morphology are capable of generating action potentials. They have resting membrane potentials of –40 to –70 mV. Under voltage clamp, the cells display inward and outward currents when stimulated by a depolarizing voltage step [Bain et al., 1995; Fraichard et al., 1995; Strübing et al., 1995]. Inward Na^+ currents inactivate rapidly and are blocked potently by the Na^+ channel antagonist tetrodotoxin. Outward K^+ currents include a delayed-rectifier-type current, which can be blocked with Cs or tetraethylammonium, and in some cells, a transient outward current that is sensitive to 4-aminopyridine. In addition, the ES cell-derived neurons express voltage-gated calcium channels. As shown by Strübing et al. [1995], distinct components of whole-cell calcium current are sensitive to specific antagonists of P-, N- and L-type channels. The ES cell-derived neurons respond to a variety of different neurotransmitters and selective receptor agonists. The inhibitory transmitters γ-aminobutyric acid and glycine activate Cl^--selective channels in most of the cells. Similarly, most of the neurons express one or more of the ionotropic glutamate receptors, which are named for the agonists amino-3-hydroxy-5-methylisoxazole-4-propionic acid (AMPA), N-methyl-*D*-aspartete (NMDA) and kainate. Moreover, distinct components of whole-cell calcium current in ES cell-derived neurons can be regulated by G protein-coupled receptors for somatostatin, γ-aminobutyric acid (GABA) and other transmitters [Strübing et al., 1997]. In all of these respects, the physiological properties of ES cell-derived neurons are similar to those of primary neurons.

The expression levels for both voltage- and ligand-gated currents, measured as current density per unit capacitance, increase with time as the cells mature [Bain et al., 1995; Strübing et al., 1995]. Voltage-gated Na^+ and K^+ currents can be detected within a day or two after EBs are plated on an adhesive substrate. Over the first 6 to 10 days their density increases to a plateau of 150–200 pA/pF. These expression levels are comparable to the density of functional ion channels observed in primary cultures of dissociated neural tissue [e.g. MacDermott and Westbrook, 1986; Nerbonne and Gurney, 1989; Wu and Barrish, 1994]. Transmitter-gated currents are first detected on the 2nd or 3rd day after plating and increase in density to reach steady levels by day 12–16. Over this same time period the cells produce abundant outgrowth of neurites, establish distinct axonal and dendritic compartments, and begin to form functional synaptic connections. Immunostaining indicates that many different neuron-specific antigens are broadly expressed in ES cells with neuronal morphology. These include NCAM, neurofilaments, class III β-tubulin, microtubule-associated proteins and synaptophysin [Bain et al., 1995; Fraichard et al., 1995; Strübing et al., 1995; Finley et al., 1996; Okabe et al., 1996].

The establishment of axons and dendrites becomes apparent when individual cells are labeled with lipophilic dyes, such as DiI (fig. 1). As is the case for primary neurons in culture and in vivo, antibodies to MAP-2 stain the dendrites, and to a lesser extent the soma, of ES cell-derived neurons, whereas antibodies to GAP-43 label axons [Finley et al., 1996]. In relatively mature cultures, 12–14 days after plating, punctate localization of synaptic vesicle antigens, including SV2, synapsin and synaptophysin, is observed at sites of presumptive intercellular transmission [Strübing et al., 1995; Finley et al., 1996; Okabe et al., 1996]. Related studies of ES cell differentiation into striated muscle have demonstrated the coaccumulation of nicotinic acetylcholine receptors with other synaptic proteins, including agrin and synaptophysin [Rohwedel et al., 1998b]. Thus, the segregation of proteins into distinct biochemical compartments is recapitulated during the in vitro differentiation of ES cells.

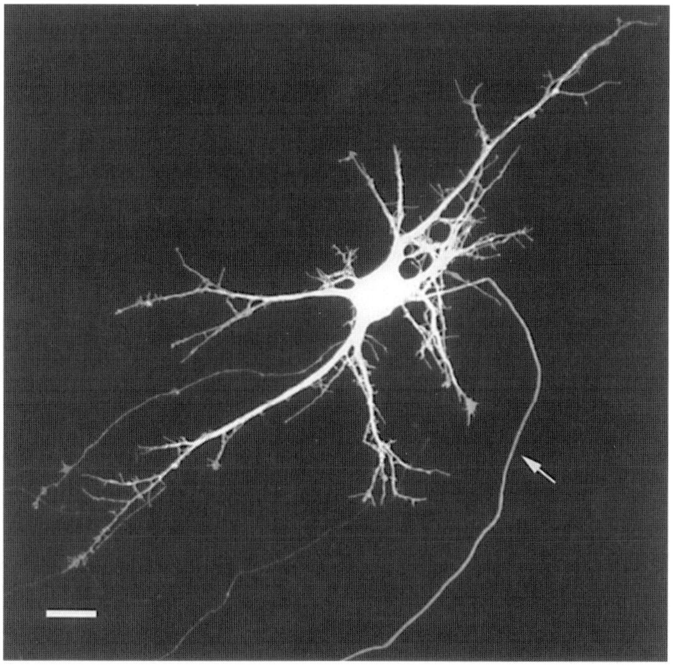

Fig. 1. Morphology of a D3 ES cell-derived neuron revealed by DiI labeling 15 days after plating. The arrow points to an axon-like fiber. Scale bar = 20 μm.

Formation of functional connections between ES cell-derived neurons can be detected as spontaneous or evoked synaptic currents [Strübing et al., 1995; Finley et al., 1996; Okabe et al., 1996]. Synaptic events are first detected between 8 and 10 days after plating. Both excitatory and inhibitory connections are observed. Studies of evoked transmission indicate that the majority of ES cell-derived neurons are excitatory (approximately 80%) while the remaining 20% are inhibitory neurons [Finley et al., 1996]. These percentages are remarkably similar to those observed in primary cultures of cerebral cortex and hippocampus [Huettner and Baughman, 1988], suggesting that the mechanisms which determine the proportion of excitatory and inhibitory cells may be similar for neurons derived from ES cells and for neurons in the brain.

Inhibitory synaptic currents in ES cell-derived neurons are blocked either by bicuculline or by strychnine, suggesting that they are mediated by GABA or glycine, respectively. Immunostaining with antibodies to glutamic acid decarboxylase (GAD) or to GABA (fig. 2) confirms that a subpopulation of ES cell-derived neurons are GABAergic.

Excitatory synaptic currents in ES cell-derived neurons are mediated by glutamate receptors. Both AMPA and

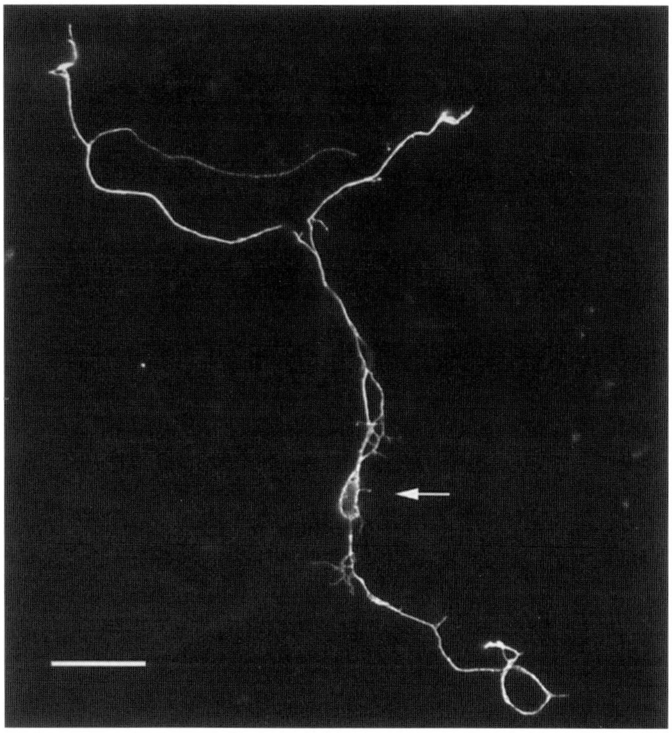

Fig. 2. GABA-like immunoreactivity in a culture of D3 ES cell-derived neurons 5 days after plating. The arrow points to the cell soma. Scale bar = 30 μm.

NMDA receptors contribute to the postsynpatic responses [Strübing et al., 1995; Finley et al., 1996], as is true for CNS neurons in vivo. Thus, electrophysiological recordings from ES-derived neurons, as well as the subcellular localization of neural antigens provide compelling evidence that the cells acquire relatively mature neuronal properties.

Glial Cells

All cultures of ES cells induced to differentiate in a neural direction contain populations of nonneuronal cells in addition to neurons. Although these have not been investigated as thoroughly as neurons, there is abundant evidence that many of them are glia. Fraichard et al. [1995] observed many examples of cells staining for glial fibrillary acidic protein or O_4 antigen, markers for astrocytes and oligodendrocytes, respectively. Bain et al. [1995] showed that glial fibrillary acidic protein mRNA is expressed in cultures of differentiated cells but not in undifferentiated cultures. Angelov et al. [1998] observed the presence of cells with astroglial, oligodendroglial and microglial marker expression. Interestingly, these appeared in a time-dependent manner that parallels their appearance during brain development. These data suggest that the in vitro system is suitable for analyzing the development and function of glial cells. The presence of glial cells also has a significant bearing on the question of how the in vitro system is related to the normal nervous system. In normal development, stem cells of the early CNS give rise to both neurons and glia. Presence of glial cells in the in vitro system strongly suggests that a similar multipotential precursor is generated in the course of ES cell neural differentiation.

Developmental Pathway from ES Cells to Neurons

The protocols reviewed above are successful at inducing ES cells to differentiate into neurons and glia. They were discovered, not by reference to normal early embryonic development, but by trial and error. This raises a very basic issue: what is the relationship of the in vitro pathway and the normal pathway of development in the embryo? There are three broad possibilities as to how the in vitro pathway could be related to normal development. The first is that there is little relation. It is imaginable that culture conditions in general, and RA in particular, turn on sets of genes resulting in neuronal differentiation in a way that bears little or no resemblance to the natural process. Normal neural development proceeds through a series of cellular progenitors that give rise to mature neurons. It is conceivable that RA directly turns on large numbers of the genes responsible for the neuronal phenotype (channels, receptors, vesicle components). This would be possible if each of these genes had a RA response element. In this model, early cellular progenitors are 'skipped' and cells go directly from ES cells to terminally differentiated neurons and glia. A second possibility is that the pathway in vitro bears substantial resemblance to the in vivo pathway but that there are also significant differences. Such differences would be imposed by culture conditions and the reduced morphological order of EBs relative to the normal embryo. Finally, it is conceivable that the pathways in vitro and in vivo are exactly the same. A priori, this is highly unlikely because it would suggest that the cellular arrangements found in the developing brain had no influence on the course of neuronal development.

At present we are in the very early stages of addressing this fundamental issue. The data we do have, however, suggests that the second alternative best describes the situation. Our reasoning is given with reference to the protocol utilized in our laboratories [Bain et al., 1995; Finley et al., 1996]. We term this the '4–/4+ protocol'; it entails culture of EBs for 4 days in the absence of RA followed by culture for 4 days in the presence of RA. There are strong parallels between events that take place in the E4–6 embryo and EBs during the first 4 days of culture. Oct-4 is a gene encoding a transcription factor that is expressed in early, pluripotent embryonic cells [Ovitt and Schöler, 1998]. Its expression is extinguished between E5 and E8 as committed lineages emerge. Oct-4 is strongly expressed in undifferentiated ES cells and its expression goes down in cultured EBs as they mature [Shen and Leder, 1992]. Thus expression in the embryo and EBs follow a similar pattern. Development of a visceral endoderm-like layer in EBs provides a second major parallel between EBs and the early embryo. The visceral endoderm is one of the earliest cell layers to emerge in the postimplantation embryo, where it grows to surround the embryo by E5. The outer layer of EBs consists of distinctive cells strongly resembling VE cells based on marker expression [Bielinska et al., 1996; Duncan et al., 1997]. In this regard the morphological appearances of the E5 embryo and EBs cultured for 4 days are strikingly similar. A third parallel relates to cavity formation. A major event in the E4–5 period of development is the formation of the proamniotic cavity via a process of cavitation of the inner cell mass. Cavitation also occurs in EBs. The parallels between cavitation in the embryo and that in EBs have been described in an elegant study [Coucouvanis and Martin, 1995]. Taken together, these results clearly demonstrate strong morphological and molecular parallels between normal embryonic development and events in EBs during the first 4 days of culture. At the time RA is added, considerable development beyond the undifferentiated ES cell has taken place.

There is also evidence based on gene expression studies that the next stages of differentiation in the EBs after the addition of RA at 4 days resemble that of the embryo [Bain et al., 1996; Li et al., 1998]. Based on these results, the conclusion that RA induces neurons by skipping the precursor stage and directly turning on late genes characteristic of terminal differentiation is virtually ruled out. Wnt-1 and MASH1 are regulatory genes expressed in early neural development within about a day of formation of the neural plate. They are not expressed in undifferentiated ES cells or EBs cultured for 4 days in the absence of RA. Addition of RA to EBs already cultured for 4 days in its absence induces the expression of wnt-1 within 2 days and MASH1 within 4 days. Markers for more advanced neural differentiation including intermediate neurofilaments and the genes for GAD (GAD65 and GAD67) are turned on after wnt-1 and MASH1. These results reveal a good correspondence between the order of gene expression in the in vitro model and the embryo. Other support for this conclusion comes from a study that uses gene targeting [Li et al.,, 1998]. A line of ES cells was created in which the SOX2 gene was targeted. The SOX2 gene codes for a DNA-binding protein selectively expressed in the early neural plate. The targeting event placed a promotorless neomycin cassette in the SOX2 gene so that only cells expressing SOX2 are neomycin-resistant. ES cells were differentiated in vitro by the 4–/4+ protocol. After neural differentiation, cultures were selected with neomycin resulting in pure cultures of neural precursor cells. Gene expression in selected cultures was analyzed. A number of regulatory genes expressed in neural precursor cells were expressed in the selected cultures, including SOX-1, pax6, pax3, mash1, and delta 1. Thus, cells undergoing neural differentiation in vitro express a number of the regulatory genes expressed in very early brain development. This supports the idea that the in vitro pathway towards mature neurons proceeds through a precursor state resembling that in vivo.

As expected, there are also important differences between in vitro ES models and normal embryonic development. We have assayed for the expression of the choline acetyltransferase gene with negative results [Bain and Gottlieb, unpubl. results]. This shows that either the pathway leading to motoneurons is not initiated or that it fails to reach the stage at which choline acetyltransferase is expressed. There is also a deviation in the pattern of wnt-1 expression [Bain et al., 1996]. In normal development, wnt-1 expression is transient. In the in vitro system, expression persists for at least 5 days postplating. Future investigations will surely reveal other differences between the CNS and the in vitro model as it stands now. These findings will provide valuable insights into the degree to which neuronal phenotypes depend on the context of an intact nervous system or are, alternatively, independent of such a context.

Neural Differentiation of ES Cells with Targeted Gene Mutations

One of the greatest potentials of this system is to analyze the effects of mutations on neurons without the need to make mice. This phase of research is just beginning.

Rohwedel et al. [1998a] showed that loss of β-integrin accelerates neuronal differentiation. The gene for GD3 synthase has been disrupted and the effects on neuronal differentiation assayed [Kawai et al., 1998]. Mutant cells lacked b-series gangliosides due to this disruption but differentiated as well as wild-type cells. While negative, the result is significant because it contradicts the established idea that this family of gangliosides is necessary for neuronal differentiation. The use of a line with a targeted mutation in the SOX2 gene [Li et al., 1998] has been described above. We anticipate a large number of studies with mutations in neural genes will be forthcoming soon.

Gene Trap Screening

Gene trap experiments with ES cells are an active area of research. In vitro differentiation of ES cells affords a way of screening for trapping of tissue-specific genes. The neuronal in vitro pathway has proven useful for identifying gene trap ES cell lines with trapping events in neural-expressed genes [Baker et al., 1997; Salminen et al., 1998; Xiang et al., 1998]. This approach will provide an array of ES cell lines with lineage- and stage-specific markers which will be useful for in vitro analysis.

Transplantation Studies

Transplantation of neurons and neural precursors into the CNS is currently an active area of research [see Brustle and McKay, 1996; Gage, 1998, for recent overviews]. These efforts are motivated by the possibility of affording treatment in cases where the brain has been damaged by traumatic injury, stroke or neurodegenerative diseases. The ES cell system has the potential of making significant contributions to transplantation research for several reasons. For example, it offers a large-scale source of cells. Cells for transplantation can be taken from any point along the entire developmental sequence from the earliest precursors to mature neurons and glia. Also, a large variety of relevant mutations obtained by gene targeting will be available, some of which may give more favorable results than normal cells. Thus far only a few transplantation studies utilizing ES cell-derived neurons and glia have been reported. Deacon et al. [1998] transplanted ES cells into the injured adult rat striatum. The transplanted ES cells engrafted well and integrated into the host striatum. Some of the transplanted cells differentiated into dopaminergic neurons while others differentiated into serotinergic neurons. Brustle et al. [1997] transplanted ES cell-derived neural precursors into the fetal rat brain. These precursors were prepared by the method of bFGF-driven expansion [Okabe et al., 1996] and transplanted into the cerebral ventricles of E16–18 rat fetuses. Results were examined within several weeks of birth. The transplanted cells differentiated into neurons, astrocytes and oligodendrocytes. Transplanted cells were integrated widely into the host CNS. We anticipate that transplantation of ES cell-derived neurons and glia will be carried out using a number of CNS injury and disease models. The availability of human ES cells [Shamblott et al., 1998; Thomson et al., 1998] adds special interest to this research area.

Species Other than Mouse

ES cells have been described in fish [Sun et al., 1995; Hong et al., 1996], chicken [Pain et al., 1996], rhesus monkey [Thomson et al., 1995, 1998c], marmoset [Thomson et al., 1996] and humans [Shamblott et al., 1998; Thomson et al., 1998a]. RA appears able to induce chicken ES cells to differentiate into neurons [Pain et al., 1996]. Neural structures figure prominently in teratomas derived from rhesus ES cells [Thomson et al., 1995, 1998b]. An interesting feature is the tendency of these cells to form organized neuroepithelia. Human ES cells have a similar propensity, forming organized neuroepithelia in teratomas created in mice [Thomson et al., 1998a] or in EBs in culture [Shamblott et al., 1998].

Unanswered Questions and Future Prospects

The investigation of ES cell-derived neural cells is in its infancy, and many basic questions about the nature of the neurons and glia derived in vitro remain unanswered. Of these many questions, two are perhaps the most basic. The first concerns the exact mechanism by which RA induces the neural pathway. Here new information at both the cellular and molecular levels is needed. At the cellular level we need to know which cells in EBs respond directly to RA. One hypothesis is that most cells in early EBs are bipotential, with the ability to form either ectodermal or mesodermal derivatives, and that RA induces the ectodermal pathway. At this stage we cannot rule out more complex models such as one in which RA acts on a small percentage of cells and these cells then provide a second type of signal to the cells which actually become neural. At the molecular level we have little mechanistic information about how RA exerts its neural-inducing action. The most firmly established mode of RA action is through the RAR and RXR families of receptors/transcription factors. These families consist of multiple genes. The number of distinct signaling proteins is much greater than the number of genes because these proteins form both homo- and heterodimers. Where in this large set of potential gene targets and regulatory activities is the actual site of RA's neural-inducing ability? Our understanding of the in vitro system will be greatly advanced when we have an answer.

Another important unanswered question concerns the phenotype of neurons generated in vitro. We already know they are polarized (that is they have axons and dendrites), make synapses and utilize glutamate, GABA or glycine as transmitters (see above). In all their properties, the ES cell-derived neurons and glia more closely resemble cells in the CNS than the peripheral nervous system. But where do they fit in the classification schemes for neurons and glia in the brain and spinal cord? For example, GABAergic neurons in the CNS are a diverse set including many distinct morphological and biochemical types. Do the GABAergic neurons in ES cell-derived cultures correspond closely to one or more of these types? Alternatively, might the ES cell-derived neurons represent a 'blend' of phenotypes of various types of GABAergic neurons? Such questions are only now beginning to be investigated. One obvious starting point will be to investigate which region-specific genes are expressed in ES cell-derived neurons. For instance, HOX gene expression is confined to the hindbrain and spinal cord. Expression of HOX genes in the model would suggest that ES cell-derived neurons had a positional identity corresponding to the caudal portion of the nervous system.

What we have learned thus far suggests that the ES cell-derived neural system accurately models a number of major aspects of development spanning the period of E4–16 in development. Thus, even in its present state, it is suitable for many types of experiments. Its great strength is that genetic, biochemical and cell biological investigations are all feasible. With future technical developments the system will become even more powerful. For instance, the approach towards generating drug-resistant cell lines via knock-ins described for the SOX2 gene [Li et al., 1998] should be general. It should be possible to make lines in which subsets of neurons can easily be selected thus generating pure cultures of say GABAergic neurons or glutaminergic neurons. This would open the door to many experiments now precluded by cellular heterogeneity. An even more ambitious goal is to direct the differentiation of ES cells along particular subpathways of neuronal differentiation. At present we know very little about the regulatory pathways responsible for this sort of choice. It is likely that many of these will be discovered. In fact the ES cell system should be useful in that effort. As we learn more, it is conceivable that we may eventually be able to start with ES cells and efficiently direct them through a series of steps so that they all end up as a single neuronal type. This goal should inspire many future efforts.

Acknowledgements

Work in the authors' labs was supported by RR12309 and the Alan A. and Edith L. Wolff Charitable Trust (D.I.G.) and NS 30888 (J.H.). We thank Dr. Michael Finley for providing figures 1 and 2. We also thank Dr. Raphael Kopan and Dr. Jeanne Nerbonne for helpful comments on the manuscript.

References

Angelov, D.N., S. Arnhold, C. Andressen, H. Grabsch, M. Puschmann, J. Hescheler, K. Addicks (1998) Temporospatial relationships between macroglia and microglia during in vitro differentiation of murine stem cells. Dev Neurosci *20*: 42–51.

Bain, G., D. Kitchens, M. Yao, J.E. Huettner, D.I. Gottlieb (1995) Embryonic stem cells express neuronal properties in vitro. Dev Biol *168:* 342–357.

Bain, G., W.J. Ray, M. Yao, D.I. Gottlieb (1996) Retinoic acid promotes neural and represses mesodermal gene expression in mouse embryonic stem cells in cutlure. Biochem Biophys Res Commun *223:* 691–694.

Baker, R.K., M.A. Haendel, B.J. Swanson, J.C. Shambaugh, B.K. Micales, G.E. Lyons (1997) In vitro preselection of gene-trapped embryonic stem cell clones for characterizing novel developmentally regulated genes in the mouse. Dev Biol *185:* 201–214.

Bielinksa, M., H. Narita, M. Heikinheimo, S.B. Porter, D.B. Wilson (1996) Erythropoiesis and vasculogenesis in embryoid bodies lacking visceral yolk sac endoderm. Blood *88:* 3720–3730.

Brustle, O., R. McKay (1996) Neuronal progenitors as tools for cell replacement in the nervous system. Curr Opin Neurobiol *6:* 688–695.

Brustle, O., A.C. Spiro, K. Karram, K. Choudhary, S. Okabe (1997) In vitro-generated neural precursors participate in mammalian brain development. Proc Natl Acad Sci USA *94:* 14809–14814.

Coucouvanis, E., G.R. Martin (1995) Signals for death and survival: A two-step mechanism for cavitation in the vertebrate embryo. Cell *83:* 279–287.

Deacon, T., J. Dinsmore, L.C. Costantini, J. Ratliff, O. Isacson (1998) Blastula-stage stem cells can differentiate into dopaminergic and serotonergic neurons after transplantation. Exp Neurol *149:* 28–41.

Duncan, S.A., A. Nagy, W. Chan (1997) Murine gastrulation requires HNF-4 regulated gene expression in the visceral endoderm: Tetraploid rescue of Hnf-4(–/–) embryos. Development *123:* 279–287.

Finley, M.F., N. Kulkarni, J.E. Huettner (1996) Synapse formation and establishment of neuronal polarity by P19 embryonic carcinoma cells and embryonic stem cells. J Neurosci *16:* 1056–1065.

Fraichard, A., O. Chassande, G. Bilbaut, C. Dehay, P. Savatier, J. Samarut (1995) In vitro differentiation of embryonic stem cells into glial cells and functional neurons. J Cell Sci *108:* 3181–3188.

Gage, F.H. (1998) Cell therapy. Nature *392:* 18–24.

Hong, Y., C. Winkler, M. Schartl (1996) Pluripotency and differentiation of embryonic stem cell lines from the medakafish *(Oryzias latipes)*. Mech Dev *60:* 33–44.

Huettner, J.E., R.W. Baughman (1988) The pharmacology of synapses formed by identified corticocollicular neurons in primary rat visual cortex. J Neurosci *8:* 160–175.

Kawai, H., K. Sango, K.A. Mullin, R.L. Proia (1998) Embryonic stem cells with a disrupted GD3 synthase gene undergo neuronal differentiation in the absence of b-series gangliosides. J Biol Chem *273:* 19634–19638.

Li, M., L. Pevny, R. Lovell-Badge, A. Smith (1998) Generation of purified neural precursors from embryonic stem cells by lineage selection. Curr Biol *8:* 971–974.

MacDermott, A.B., G.L. Westbrook (1986) Early development of voltage-dependent sodium currents in cultured mouse spinal cord neurons. Dev Biol *113:* 317–326.

Martin, G. (1981) Isolation of a pluripotent cell line from early mouse embryos cultured in medium conditioned by teratocarcinoma stem cells. Proc Natl Acad Sci USA *78:* 7634–7638.

Nerbonne, J.M., A.M. Gurney (1989) Development of excitable membrane properties in mammalian sympathetic neurons. J Neurosci *9:* 3272–3286.

Okabe, S., K. Forsberg-Nilsson, A.C. Spiro, M. Segal, R.D. McKay (1996) Development of neuronal precursor cells and functional postmitotic neurons from embryonic stem cells in vitro. Mech Dev *59:* 89–102.

Ovitt, C.E., H.R. Schöler (1998) The molecular biology of Oct-4 in the early mouse embryo. Mol Hum Reprod *4:* 1021–1031.

Pain, B., M.E. Clark, M. Shen, H. Nakazawa, M. Sakurai, J. Samarut, R.J. Etches (1996) Long-term in vitro culture and characterisation of avian embryonic stem cells with multiple morphogenetic potentialities. Development *122:* 2339–2348.

Rohwedel, J., K. Guan, W. Zuschratter, S. Jin, G. Ahnert-Hilger, D. Fürst, R. Fässler, A.M. Wobus (1998a) Loss of beta integrin function results in a retardation of myogenic, but an acceleration of neuronal, differentiation of embryonic stem cells in vitro. Dev Biol *201:* 167–184.

Rohwedel, J., T. Kleppisch, U. Pich, K. Guan, S. Jin, W. Zuschratter, C. Hopf, W. Hoch, J. Hescheler, V. Witzemann, A.M. Wobus (1998b) Formation of postsynaptic-like membranes during differentiation of embryonic stem cells in vitro. Exp Cell Res *239:* 214–225.

Salminen, M., B.I. Meyer, P. Gruss (1998) Efficient poly A trap approach allows the capture of genes specifically active in differentiated embryonic stem cells and in mouse embryos. Dev Dyn *212:* 326–333.

Shamblott, M., J. Axelman, S. Wang, E. Bugg, J. Littlefield, P. Donovan, P. Blumenthal, G. Huggins, J. Gearhart (1998) Derivation of pluripotent stem cells from cultured human primordial germ cells. Proc Natl Acad Sci USA *95:* 13726–13731.

Shen, M., P. Leder (1992) Leukemia inhibitory factor is expressed by the preimplantation uterus and selectively blocks primitive ectoderm formation in vitro. Proc Natl Acad Sci USA *89:* 8240–8244.

Strübing, C., G. Ahnert-Hilger, J. Shan, B. Wiedenmann, J. Hescheler, A.M. Wobus (1995) Differentiation of pluripotent embryonic stem cells into the neuronal lineage in vitro gives rise to mature inhibitory and excitatory neurons. Mech Dev *53:* 275–287.

Strübing, C., J. Rohwedel, G. Ahnerthilger, B. Wiedenmann, J. Hescheler, A.M. Wobus (1997) Development of G protein-mediated Ca^{++} channel regulation in mouse embryonic stem cell-derived neurons. Eur J Neurosci *9:* 824–832.

Sun, L., C.S. Bradford, C. Ghosh, P. Collodi, D.W. Barnes (1995) ES-like cell cultures derived from early zebrafish embryos. Mol Mar Biol Biotechnol *4:* 193–199.

Thomson, J.A., J. Kalishman, T.G. Golos, M. Durning, C.P. Harris, R.A. Becker, J.P. Hearn (1995) Isolation of a primate embryonic stem cell line. Proc Natl Acad Sci USA *92:* 7844–7848.

Thomson, J.A., J. Itskovitz-Eldor, S. Shapiro, M. Waknitz, J. Swiergiel, V. Marshall, J. Jones (1998a) Embryonic stem cell lines derived from human blastocytes. Science *282:* 1145–1147.

Thomson, J.A., V.S. Marshall, J.Q. Trojanowski (1998b) Neural differentiation of rhesus embryonic stem cells. APMIS *106:* 149–156.

Thomson, J.A., V.S. Marshall (1998c) Primate embryonic stem cells. Curr Top Dev Biol *38:* 133–165.

Thomson, J.A., J. Kalishman, T.G. Golos, M. Durning, C.P. Harris, J.P. Hearn (1996) Pluripotent cell lines derived from common marmoset *(Callithrix jacchus)* blastocysts. Biol Reprod *55:* 254–259.

Wu, R.L., M.E. Barrish (1994) Astroglial modulation of transient potassium current development in cultured mouse hippocampal neurons. J Neurosci *14:* 1677–1687.

Xiang, J.W., R. Battaglino, A. Leahy, H. Stuhlmann (1998) Large-scale screening for developmental genes in embryonic stem cells and embryoid bodies using retroviral entrapment vectors. Dev Dyn *212:* 181–197.

Cells
Tissues
Organs

Cells Tissues Organs 1999;165:173–180

Accepted after revision: March 14, 1999

Embryonic Stem Cell-Derived Adipogenesis

Christian Dani

Laboratoire de Biologie du Développement du Tissu Adipeux, Centre de Biochimie (UMR 6543 CNRS), Nice, France

Key Words
Embryonic stem cells · Adipocytes

Abstract
Key events leading to terminal differentiation of preadipocytes into adipocytes have been characterized in the recent years. However, master genes that commit progression from multipotent mesenchymal stem cell to the adipoblast stage of development have not yet been identified. The use of embryonic stem (ES) cells as a route to study early events in adipogenesis and to characterize factors involved in the decision of stem cells to follow the adipogenic pathway is described in this paper. The capacity of lif$^{-/-}$ and lifr$^{-/-}$ ES cells to undergo adipocyte differentiation is reported as an application of mutant ES cells to study gene function during the development of adipose cells.

Introduction

Two kinds of adipose tissues with specific functions are present in rodents and humans. The white adipose tissue (WAT) stores energy in the form of triglycerides in times of nutritional excess and releases free fatty acids during food deprivation. WAT mass is determined by the balance between energy intake and expenditure. Alterations of this steady state can lead to overweight and obesity often accompanied by metabolic disorders associated with cardiovascular diseases (hypertension, type II diabetes). The brown adipose tissue (BAT) is the major site of compensatory thermogenesis to maintain body temperature. The thermogenic activity of BAT is explained by the activity of uncoupled mitochondria which dissipate a large proportion of energy as heat when metabolizing free fatty acids to produce ATP. The uncoupled protein-1

Abbreviations used in this paper

A_2COL_6	α chain 2 of type VI collagen
a-FABP	adipocyte-fatty acid binding protein
BAT	brown adipose tissue
C/EBPs	CCAAT/enhancer binding proteins
EB	embryoid body
ES	embryonic stem cells
LIF	leukemia inhibitory factor
LIF-R	leukemia inhibitory factor receptor
PGC-1	PPAR gamma coactivator-1
PPARs	peroxisome proliferative-activated receptors
RA	retinoic acid
RARs	retinoic acid receptors
RXRs	retinoid acid receptors
UCP-1	uncoupled protein-1
WAT	white adipose tissue

KARGER

Fax + 41 61 306 12 34
E-Mail karger@karger.ch
www.karger.com

1422–6405/99/1654–0173$17.50/0

Accessible online at:
www.karger.com/journals/cto

Dr. Christian Dani
Laboratoire de Biologie du Développement du Tissu Adipeux
Centre de Biochimie (UMR 6543 CNRS), UNSA, Faculté des Sciences
Parc Valrose, F–06108 Nice Cedex 2 (France)
Tel. +33 4 92 07 6436, Fax +33 4 92 07 6404, E-Mail dani@unice.fr

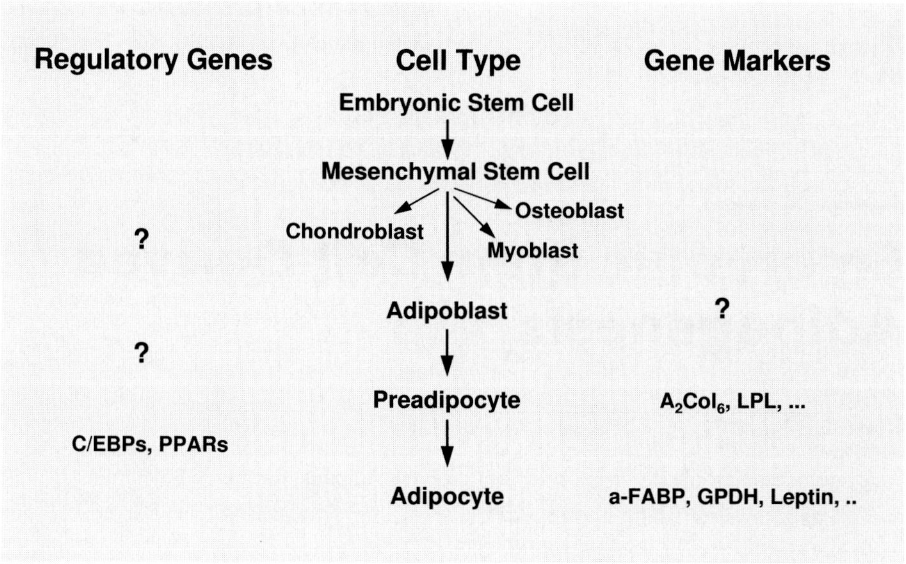

Fig. 1. Adipose cell differentiation is a sequential process. A_2Col_6 and lipoprotein lipase (LPL) are early markers of differentiation. a-FABP, glycerol-phosphate dehydrogenase (GPDH) and leptin are late markers of differentiation. C/EBPs α, β and δ and PPARs δ and γ are transcriptional factors playing a key role in the terminal differentiation process.

(UCP-1) is expressed exclusively in mitochondria of BAT and is responsible for the thermogenic activity of brown adipocytes [Ricquier and Bouillaud, 1997]. In humans, adipose tissue (most likely BAT) develops between the second and third trimester of gestation. WAT develops mainly in neonates and adolescents but also in adults whereas BAT almost disappears after birth. In rodents, BAT develops late during fetal life and is identifiable at birth. In contrast, WAT cannot be detected mascroscopically during embryogenesis. Adipose tissue can be identified only when it contains adipocytes (i.e., terminally differentiated cells). Because there are no specific markers that can be used to date to identify adipoblasts and preadipocytes in vivo, very little is known about the ontogeny of adipose tissue. The capacity of embryonic stem (ES) cells to undergo adipocyte differentiation in vitro provides a promising model for studying early differentiative events in adipogenesis and for identifying regulatory genes involved in the commitment of multipotent mesenchymal stem cell to the adipoblast lineage.

In vitro Differentiation of Preadipocytes into Adipocytes

The development of established preadipocyte cell lines has greatly facilitated the study of the different steps leading to terminal differentiation of preadipocytes into adipocytes. Several cell lines are used for in vitro study of the process of differentiation and basically they display the same sequence of events [Ailhaud et al., 1992; Hwang et al., 1997]. Adipose cell differentiation is a multistep process characterized by a sequence of events during which adipoblasts divide until confluence (fig. 1). Adipoblasts, derived from a mesenchymal precursor, are fibroblast-like cells. No specific markers of that stage have been identified so far. Growth arrest leads to the emergence of early markers of differentiation, such as lipoprotein lipase and the α chain 2 of type VI collagen (A_2COL_6), and to the formation of preadipose cells that have not yet accumulated lipids. Finally, terminal differentiation of preadipocytes into adipocytes leads to the emergence of a very large set of proteins involved in lipid accumulation and mobilization, resulting in the formation of lipid-containing cells. Mature adipocytes with lipid droplets can be morphologically identified both in vitro and in vivo. Several hormones and growth factors as well as arachidonic acid and some of its metabolites are able to modulate adipocyte differentiation. The responsiveness of preadipose cells to external signals may vary according to culture conditions tested and differences in the stage of development at which clonal lines have been established [Smas and Sul, 1995].

Key events of terminal differentiation have been identified in recent years by the characterization of transcriptional factors playing a regulatory role in the differentiation process. The best characterized factors identified as important in the development of mature adipocytes are members of the CCAAT/enhancer binding proteins (C/EBPs) and peroxisome proliferative-activated receptors (PPARs) families. C/EBPs are basic leucine zipper proteins and PPARs belong to the nuclear hormone receptor superfamily. Several lines of evidence indicate that PPARγ, which is more abundant in adipocytes than in other cell types, is a trigger of terminal differentiation [Tontonoz et al., 1994, 1995]. C/EBPs expression is not restricted to adipose tissue. However, the generation by homologous recombination of C/EBPα-negative mice [Wang et al., 1995] or C/EBPβ/δ-negative mice has established an essential role of these factors for the acquisition of adipocytes both in vitro and in vivo. It is now clear that these transcriptional factors play a sequential role in the adipocyte differentiation process and expression of C/EBPβ and C/EBPδ is so far the earliest to occur in preadipose cells. However, master genes that commit progression from the postulated multipotent mesenchymal stem cell to the adipoblast stage of development as well as factors required for adipoblast stem cell renewal have not yet been identified.

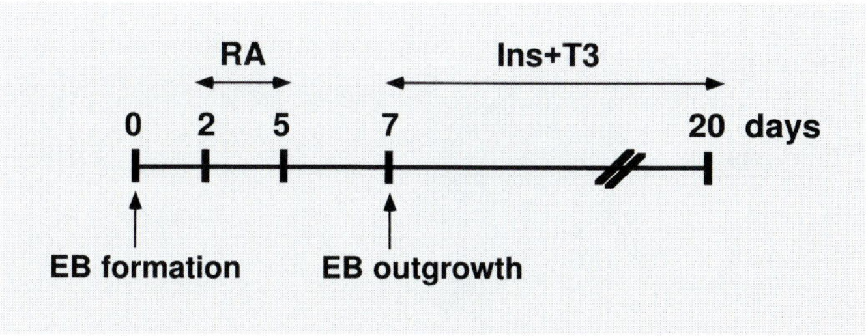

Fig. 2. Experimental protocol used for the commitment of ES cells into the adipocyte lineage. The permissive period for commitment into the adipocyte is influenced by RA whereas terminal adipocyte differentiation takes place in the presence of 85 n*M* insulin (Ins) and 2 n*M* triiodothyronine (T_3).

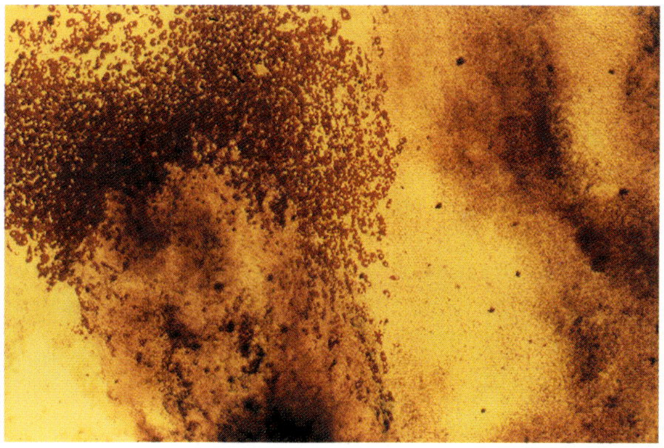

Fig. 3. EB outgrowth-derived adipocytes. A 20-day outgrowth stained with oil red O for fat droplets. Adipocytes which developed out of the center of the outgrowth are shown.

In vitro Differentiation of ES Cells along an Adipogenic Pathway: Critical Role of All-*trans*-Retinoic Acid

ES cells have previously been shown to differentiate spontaneously into various lineages in culture [Doetschman et al., 1985]. The first morphological observation of adipocyte-like cells derived from ES cells was reported by Fried et al. [1992]. However, the number of ES cell-derived adipocytes was low because spontaneous commitment of ES cells into the adipocyte lineage is rare. In order to use ES cells to study adipocyte development it was therefore essential to determine conditions of culture that favor the adipocyte differentiation at a high rate and to better characterize ES cell-derived adipocyte-like cells. This step has been achieved recently [Dani et al., 1997]. A prerequisite for the commitment of ES cells into the adipocyte lineage is to treat ES cell-derived embryoid bodies at an early stage of their differentiation with retinoic acid (RA) for a short period of time. Two phases can be distinguished in the development of adipogenesis from ES cells (fig. 2): the first phase, between day 2 and 5 after embryoid body (EB) formation, corresponds to a permissive period for the commitment of ES cells which is influenced by all-*trans*-RA. The second phase corresponds to the permissive period for terminal differentiation and requires adipogenic hormones as previously shown for the differentiation of cells from preadipose clonal lines. The treatment leads to 50–70% of outgrowths containing adipose cells compared to 2–5% in the absence of RA treatment. Figure 3 shows a 20-day-old outgrowth derived from an RA-treated EB: a large adipocyte colony which developed out of the center of the body can be distinguished.

RA could not be substituted by hormones or compounds known to be important for terminal differentiation. Treatment of early EB with either insulin, triiodothyronine, dexamethasone or potent activators of PPARs, such as the thiazolidinedione BRL49653 and the nonmetabolizable fatty acid 2-bromopalmitate, alone or in combination, led to a low level of adipogenesis (5%). Among factors which have previously been reported to modulate terminal adipocyte differentiation, RA is the only naturally occurring compound able to trigger development of adipose cells from ES cells.

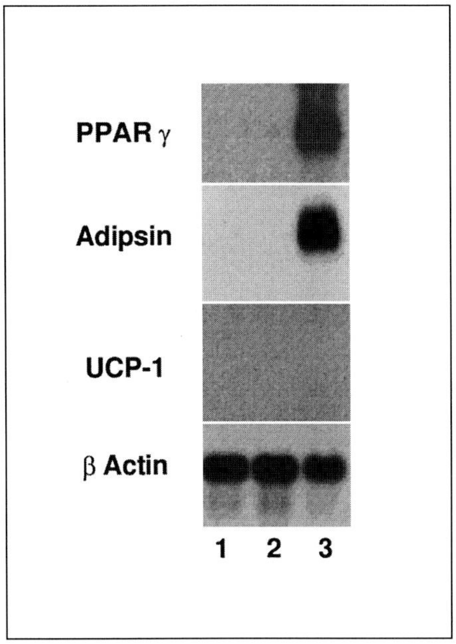

Fig. 4. RA-treated EB outgrowths express white adipocyte-specific markers. Poly(A)$^+$ RNA (3 μg) were prepared from undifferentiated ES cells (1) or from 20-day-old EBs pretreated with 0.1% DMSO (2) or 10^{-8} *M* RA (3). RA pretreatment was required for commitment of ES cells into the adipogenic pathway and no adipocyte-specific markers were detected in untreated or 0.1% ethanol-pretreated EBs.

As mentioned in the introduction, the main function of adipocytes as energy source is to store triglycerides (lipogenic activity) and to release free fatty acids (lipolytic activity) upon hormonal conditions. It has been shown that EB-derived adipocytes display both lipogenic and lipolytic activities in response to insulin and to β-adrenergic agonists, respectively [Dani et al., 1997], indicating that mature and functional adipocytes are indeed formed from ES cells in vitro.

ES Cell-Derived Adipocytes Exhibit Specific Markers of White Adipocytes but Not of Brown Adipocytes

In rodents, WAT and BAT are located in different depots. WAT is mainly present in subcutaneous and visceral areas whereas BAT is located in the interscapular area. However, a low number of brown adipocytes can be detected even in WAT [Young et al., 1984; Cousin et al., 1992]. A conversion of brown adipocytes into white adipocytes during development and in pathological situations has been suggested and the hypothesis of a cellular plasticity of adipose tissue had been proposed some years ago [Casteilla et al., 1994]. The existence of a common or distinct adipoblast precursor for white and brown adipocytes remains controversial. As shown in figure 4, treatment of EBs with 10^{-8} *M* RA during the determination period of time led subsequently to a high expression of adipocyte-specific genes, such as PPARγ and adipsin. In contrast, expression of UCP-1 gene, a specific marker of brown adipocytes, could not be detected. A reverse transcription-PCR analysis confirmed that the UCP-1 gene was not expressed in 20-day-old outgrowths. The absence of UCP-1 RNA in differentiating EBs suggests either that (1) RA triggers commitment of ES cells only towards white adipocytes without any plasticity potential, which is in favor of the distinct precursor hypothesis, or (2) brown adipocyte lineage is also generated but *trans*-acting factors required to express UCP-1 gene are missing. Cloning of a novel transcription coactivator of nuclear receptors, named PPAR gamma coactivator (PGC)-1, which is highly expressed in brown fat has recently been reported [Puigserver et al., 1998]. Interestingly, PGC-1 is not expressed in white adipocytes. PGC-1 seems to be a key regulator of UCP-1 gene expression and of mitochondriogenesis as its ectopic expression in white adipocytes turns them into brown adipocyte-like cells [Puigserver et al., 1998]. Studies of the expression of PGC-1 in differentiating EBs and genetic manipulations of ES cells may shed new light on the relationship between brown and white adipoblast precursors.

Expression of Key Regulators of Terminal Differentiation during Development of ES Cell-Derived EB

PPARs (PPARδ and PPARγ) and C/EBPs (C/EBPβ, C/EPBδ and C/EBPα) are nuclear factors that regulate genes involved in lipid metabolism. C/EBPα seems to be important to maintain the adipocyte differentiated phenotype, whereas several lines of evidence indicate that PPARs and C/EBPβ and C/EBPδ are triggers of terminal differentiation of preadipocytes into adipocytes. The role of these factors in the commitment of stem cells into the adipocyte lineage has been addressed by studying their expression during the determination and the differentiation periods of ES cells. As shown in figure 5, expression of PPARγ and C/EBPβ is low during the determination phase and parallels expression of adipocyte-fatty acid binding protein (a-FABP) which is a marker of terminal differentiation. This result suggests that PPARγ and C/

EBPβ are not regulatory genes for the commitment of ES cells into the adipocyte lineage. It has previously been reported that PPARδ gene expression is detected early during rat embryonic development and preceded expression of PPARγ [Braissant and Wahli, 1998]. The same temporal pattern of expression is conserved in developing EBs. In contrast to PPARγ, PPARδ is strongly expressed during the determination phase of ES cells suggesting that this factor could be a good candidate as master gene involved in the commitment of mesenchymal precursors into the adipocyte lineage. However, expression of PPARδ gene is not restricted to adipose tissue and its expression is not modified by the treatment required to induce adipogenesis of ES cells. Stimulation of early EBs by potent activators of PPARδ such as fatty acid 2-bromoplamitate or carbacyclin could not trigger differentiation of EBs along an adipogenic pathway. Altogether, these results suggest that C/EBPβ and PPARs are not involved in the early events leading to the development of adipose cells. The generation of ES cells deficient for PPARδ and/or PPARγ will facilitate elucidation of the role of these transcription factors during the different stages of adipogenesis. The differentiation culture system combined with genetic manipulations of undifferentiated ES cells, such as gene trapping and gain or loss of function, should provide a means to identify novel regulatory genes involved in early determinative events in adipogenesis.

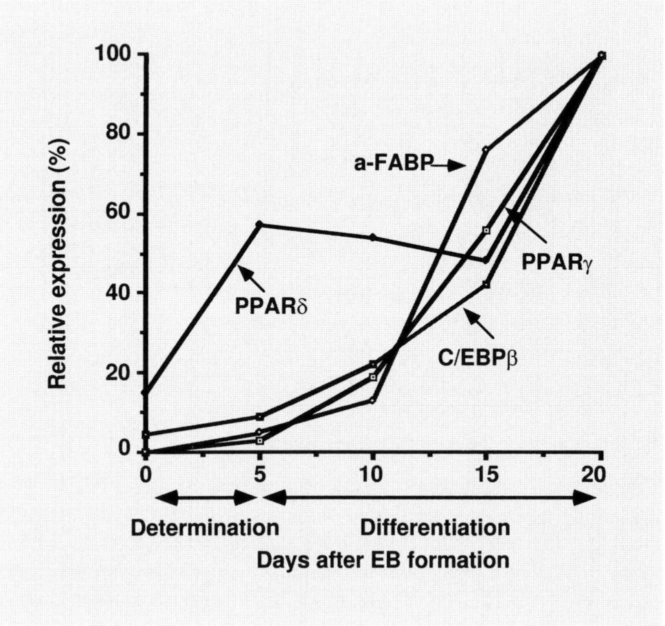

Fig. 5. Kinetics of gene expression in developing EBs. EBs were treated from day 2 to day 5 with 10^{-8} *M* RA (determination period of time) and plated in differentiation medium for 15 days (differentiation period of time). Total RNA was prepared at indicated times and analyzed for expression of indicated genes by Northern blotting. Quantification of the hybridization signal was performed using a PhosphorImager apparatus. The results are expressed by taking as 100% the maximal signal obtained for each probe. The results are representative of two independent experiments.

Application of Mutant ES Cells to Study Gene Function during the Development of Adipose Cells

We have recently observed that leukemia inhibitory factor (LIF) and LIF receptor (LIF-R) genes are developmentally regulated during the differentiation of preadipocytes to adipocytes [Aubert et al., unpubl. results]. The fact that LIF and LIF-R are both expressed during the first step of adipocyte differentiation led us to speculate that this pathway plays a regulatory role in adipogenesis. The role of LIF was addressed by investigating whether lif$^{-/-}$ ES cells were able to undergo adipocyte differentiation. These mutant ES cells have been generated by gene targeting via two rounds of homologous recombination [Dani et al., 1998]. The results shown in figure 6 indicate that LIF-null ES cells underwent adipogenesis with comparable efficiency to wild-type cells, which is in agreement with studies of LIF mutant mice indicating that a lack of LIF expression did not prevent the development of adipose tissue [Dani et al., unpubl. results]. LIF belongs to

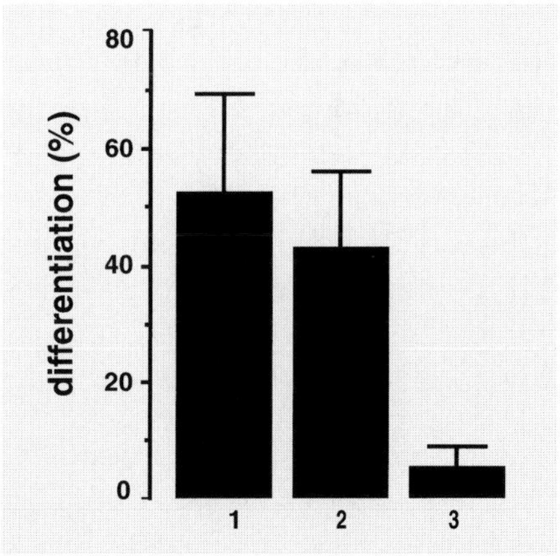

Fig. 6. Capacity of LIF- and LIF-R-deficient ES cells to undergo adipocyte differentiation. EB derived from wt (1), lif$^{-/-}$ (2) and lifr$^{-/-}$ ES cells (3) were induced to undergo adipocyte differentiation as described in figure 2. Twenty days after EB formation the percentage of outgrowths showing adipocyte colonies was scored. Number of outgrowths (no), number of independent clones (ncl) checked and number of experiments (ne) performed are, respectively: (1) no = 334, ncl = 3, ne = 4; (2) no = 140, ncl = 1, ne = 2; (3) no = 609, ncl = 2, ne = 3.

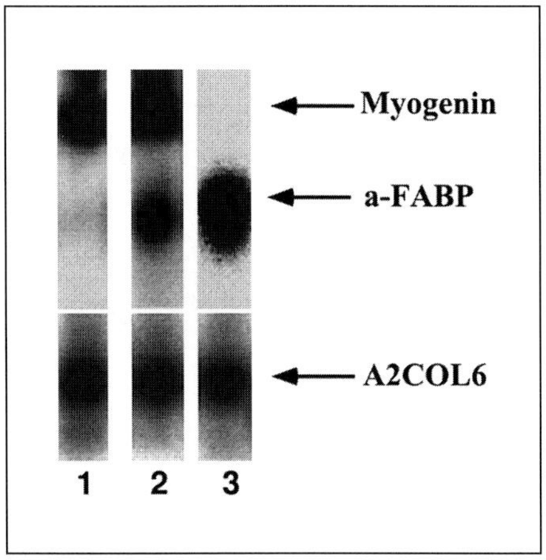

Fig. 7. PCR analysis of expression of myogenin, a-FABP and A_2COL_6 genes in a single 20-day-old outgrowth. EBs were pretreated from day 2 to day 5 with 0.1% DMSO (1), or 10^{-8} *M* RA (2), or 10^{-7} *M* RA (3). RNA from individual outgrowth was prepared as previously described [Dani et al., 1997] and transcripts for indicated genes were amplified using the following pairs of primers: a-FABP, 5′ GATGCCTTTGTGGGAACCTGG 3′ and 5′ TTCATCGAATTC-CACGCCCAG 3′; myogenin, 5′ AGCTCCCTCAACCAGGAGGA 3′ and 5′ GGGCTCTCTGGACTCCATCT 3′; A_2COL_6, 5′ AACTT-CGCCGTGGTCATCACTGACG 3′ and 5′ AGGAATCTCCAGG-CAGCTCACCTTG 3′. Size of the amplified cDNA products are 213, 500 and 300 bp, respectively. Size of genomic DNA-derived bands are 2.4, 1.5 and 0.5 kbp, respectively. Samples (1/10 of the assay) were analyzed by electrophoresis in 2% agarose. The products were blotted onto a Hybond membrane and hybridized with appropriate cDNA probes.

the IL-6 cytokine family and a feature of members of this family is the redundancy of biological functions [Taga and Kishimoto, 1997]. Therefore, we have postulated that LIF-related cytokines could compensate for the lack of LIF both in vivo and in vitro. The role of LIF-R during adipogenesis was therefore investigated. Homozygous LIF-R mutant mice are not viable [Li et al., 1995; Ware et al., 1995], precluding the study of the role of this signaling pathway during the development of adipose tissue in vivo. However, $lifr^{-/-}$ ES cells have been generated by Li in Smith's laboratory (Center for Genome Research, Edinburgh, Scotland) and we have shown that the capacity of LIF-R null ES cells to undergo adipocyte differentiation is dramatically reduced. Only 5–7% of outgrowths derived from mutant cells contained adipocyte colonies compared to 55–70% of outgrowths derived from wild-type ES cells [Dani et al., unpubl. results].

The use of genetically modified ES cells combined with conditions of culture to commit stem cells into the adipogenic pathway has allowed us to provide evidence that LIF-R plays a critical role in the development of adipose cells.

Differentiation of ES Cells into Adipocytes Inhibits the Development of the Skeletal Myocyte Lineage

Adipocytes and skeletal myocytes are believed to be derived from the same mesenchymal stem cell precursor [Taylor and Jones, 1979] and it has been suggested that in vitro the skeletal muscle and adipose development programs are mutually exclusive [Weintraub et al., 1989; Hu et al., 1995; Teboul et al., 1995]. In vivo there is often an inverse relationship between skeletal muscle and adipose tissue development. This phenomenon can be observed in some pathologies such as in Duchenne muscular dystrophy [Anderson and Kunkel, 1992]. In contrast to the adipocyte lineage, the skeletal myocyte lineage appears spontaneously during differentiation of ES cells [Doetschman et al., 1985; Rohwedel et al., 1994]. As shown in figure 7, single EB pretreated with a low concentration of RA (10^{-8} *M*) can give rise subsequently to both adipocytes and skeletal myocytes (determined by expression of a-FABP and myogenin genes, respectively). However, as the concentration of RA is increased, a shift in the progression of the differentiation program occurs [Pobre and Dani, unpubl. results]. At an RA concentration higher than 10^{-8} *M*, the expression of myogenin was inhibited and expression of a-FABP was increased (fig. 7). Expression of a-FABP and myogenin genes was paralleled by the development of adipocytes and myocytes scored by microscopic examination. Expression of the A_2COL_6 gene, which is mainly expressed by mesenchymal cells [Ibrahimi et al., 1993; Marvulli et al., 1996], was not modified suggesting that pretreatment of early EBs with RA did not lead to generalized changes in the development program of ES cells. Altogether, these results indicate that a switch from myogenesis to adipogenesis can be induced by RA in a concentration-dependent manner. Although studies of expression of early gene markers of skeletal myogenesis, such as Myf 5 or MyoD, are required to know at which stage the development of myoblast precursors is blocked, these results strongly suggest that the permissive period for the commitment of ES cells into the adipocyte lineage is also critical for the myocyte lineage. In vitro differentiation of ES cells could allow characterization of factors involved in the decision of stem cells to follow the adipogenic or myogenic developmental pathway.

Conclusion and Future Trends

In vitro differentiation of ES cells into adipocytes will facilitate both the determination of the role of genes during the process of adipose cell differentiation and the identification of novel genes, the expression of which is required for commitment of stem cells into the adipocyte lineage. One possible route to identify master genes for adipocyte commitment is to identify genes regulated by RA in early EB. The effects of RA are mediated through specific receptors which belong to the nuclear receptor family of ligand-dependent transcription factors. The signaling pathway activated by RA is complex. Indeed, there are two distinct subfamilies of retinoic acid receptors. The retinoic acid receptors (RARs) and the retinoid acid receptors (RXRs), each have three members, α, β and γ, which can form homodimers and heterodimers. The three different isotypes of RARs and RXRs are encoded by different genes displaying a characteristic developmental and tissue-specific regulated expression [Gudas, 1994; Chambon, 1996]. Generation of F9 murine embryonal carcinoma cells deficient for either RXRα or RARs (α or γ) or RXRα and RARs led to the conclusion that the heterodimers can exert both specific and redundant functions in regulating gene expression and cell differentiation [Chiba et al., 1997]. In preadipose clonal lines low concentrations of RA stimulate terminal adipocyte differentiation whereas high concentrations of RA inhibit terminal differentiation of preadipocytes into adipocytes [Safonova et al., 1994]. RARα seems to mediate this inhibitory effect on terminal differentiation [Kamei et al., 1994]. It would be interesting to identify the isotype of RAR (RARα, β or γ) involved in the commitment of stem cells along the adipogenic pathway. The use of synthetic retinoids selective for either RARα, RARβ or RARγ allowed Roy et al. [1995] to investigate the effects of RAR isotypes on the morphological differentiation of F9 cells into primitive and parietal endoderm. This approach could be equally performed on ES cells to determine whether activation of a specific RAR or RAR/RXR complex is sufficient to trigger EBs into the adipocyte lineage. Next, characterization of genes regulated by the RAR(s) of interest during the critical window of time for adipogenic commitment may lead to the identification of regulatory events involved in early development of adipose cells.

Acknowledgements

The author is grateful to Professor R. Negrel, Professor G. Ailhaud and all members of the laboratory of 'Biology of Adipose Tissue Development'. The author also thanks G. Oillaux for secretarial assistance, Dr. L. Casteilla for helpful discussions on the plasticity of adipose tissues and Dr. A. Smith for helpful discussions and critical reading of the manuscript.

References

Ailhaud, G., P. Grimaldi, R. Négrel (1992) Cellular and molecular aspects of adipose tissue development. Annu Rev Nutr *12:* 207–233.

Anderson, M.S., L.M. Kunkel (1992) The molecular and biochemical basis of Duchenne muscular dystrophy. Trends Biochem Sci *17:* 289–292.

Braissant, O., W. Wahli (1998) Differential expression of peroxisome proliferator-activated receptor-α, -β, and -γ during rat embryonic development. Encocrinology *139:* 2748–2754.

Casteilla, L., B. Cousin, N. Viguerie-Bascands, D. Larrouy, L. Pénicaud (1994) Hétérogénéité et plasticité cellulaires des tissus adipeux. Médecine Sciences *10:* 1099–1106.

Chambon, P. (1996) A decade of molecular biology of retinoic acid receptors. FASEB J *10:* 940–954.

Chiba, H., J. Clifford, D. Metzger, P. Chambon (1997) Specific and redundant functions of retinoid X receptor/retinoic acid receptor heterodimers in differentiation, proliferation, and apoptosis of F9 embryonal carcinoma cells. J Cell Biol *139:* 735–747.

Cousin, B., S. Cinti, M. Morroni, S. Raimbault, D. Pénicaud, L. Casteilla (1992) Occurrence of brown adipocytes in rat white adipose tissue: Molecular and morphological characterization. J Cell Sci *103:* 931–942.

Dani, C., I. Chambers, S. Johnstone, M. Robertson, B. Ebrahimi, M. Saito, T. Taga, M. Li, T. Burdon, J. Nichols, A. Smith (1998) Paracrine induction of stem cell renewal by LIF-deficient cells: A new ES cell regulatory pathway. Dev Biol *203:* 149–162.

Dani, C., A.G. Smith, S. Dessolin, P. Leroy, L. Staccini, P. Villageois, C. Darimont, G. Ailhaud (1997) Differentiation of embryonic stem cells into adipocytes in vitro. J Cell Sci *110:* 1279–1285.

Doetschman, T.C., H. Eistetter, M. Katz, W. Schmidt, R. Kemler (1985) The in vitro development of blastocyte-derived embryonic stem cell lines: Formation of visceral yolk sac, blood islands and myocardium. J Embryol Exp Morphol *87:* 27–45.

Fried, S.J., R.S. Johnson, R.M. Mortensen, V.E. Papaioannou, B.M. Spiegelman, M.E. Greenberg (1992) Growth and differentiation of embryonic stem cells that lack an intact *c-fos* gene. Proc Natl Acad Sci USA *89:* 9306–9310.

Gudas, L.J. (1994) Retinoids and vertebrate development. J Biol Chem *269:* 15399–15402.

Hu, E., P. Tontonoz, B.M. Spiegelman (1995) Transdifferentiation of myoblasts by the adipogenic transcription factors PPARγ and C/EBPα. Proc Natl Acad Sci USA *92:* 9856–9860.

Hwang, C.S., T.M. Loftus, S. Mandrup, M.D. Lane (1997) Adipocyte differentiation and leptin expression. Annu Rev Cell Dev Biol *13:* 231–259.

Ibrahimi, A., B. Bertrand, S. Bardon, E.Z. Amri, P. Grimaldi, G. Ailhaud, C. Dani (1993) Cloning of α2 chain of type VI collagen and expression during mouse development. Biochem J *289:* 141–147.

Kamei, Y., T. Kawada, J. Mizukami, E. Sugimoto (1994) The prevention of adipose differentiation of 3T3-L1 cells caused by retinoic acid is elicited through retinoic acid receptor alpha. Life Sci *55:* 307–312.

Li, M., M. Sendtner, A. Smith (1995) Essential function of LIF receptor in motor neurons. Nature *378:* 724–727.

Marvulli, D., D. Volpin, G.M. Bressan (1996) Spatial and temporal changes of type VI collagen expression during mouse development. Dev Dyn *206:* 447–454.

Puigserver, P., Z. Wu, C.W. Parl, R. Graves, M. Wright, B.M. Spiegelman (1998) A cold-inducible coactivator of nuclear receptors linked to adaptive thermogenesis. Cell *92:* 829–839.

Ricquier, D., F. Bouillaud (1997) The mitochondrial uncoupling protein: Structural and genetic studies. Prog Nucleic Acid Res Mol Biol *56:* 83–108.

Rohwedel, J., V. Maltsev, E. Bober, H.H. Arnold, J. Hescheler, A.M. Wobus (1994) Muscle cell differentiation of embryonic stem cells reflects myogenesis in vivo: Developmentally regulated expression of myogenic determination genes and functional expression of ionic currents. Dev Biol *164:* 87–101.

Roy, B., R. Taneja, P. Chambon (1995) Synergistic activation of retinoic acid (RA)-responsive genes and induction of embryonal carcinoma cell differentiation by an RA receptor α (RARα)-, RARβ-, or RARγ-selective ligand in combination with a retinoid X receptor-specific ligand. Mol Cell Biol *15:* 6481–6487.

Safonova, I., C. Darimont, E.Z. Amri, P. Grimaldi, G. Ailhaud, U. Reichert, B. Shroot (1994) Retinoids are positive effectors of adipose cell differentiation. Mol Cell Endocrinol *101:* 201–211.

Smas, C.M., H.S. Sul (1995) Control of adipocyte differentiation. Biochem J *309:* 697–710.

Taga, T., T. Kishimoto (1997) gp 130 and the interleukin-6 family of cytokines. Annu Rev Immunol *15:* 797–819.

Tanaka, T., N. Yoshida, T. Kishimoto, S. Akira (1997) Defective adipocyte differentiation in mice lacking the C/EBPβ and/or C/EBPδ gene. EMBO J *16:* 7432–7443.

Taylor, S.M., P.A. Jones (1979) Multiple new phenotypes induced in 10T1/2 and 3T3 cells treated with 5-azacytidine. Cell *17:* 771–779.

Teboul, L., D. Gaillard, L. Staccini, H. Inadera, E.Z. Amri, P.A. Grimaldi (1995) Thiazolidinediones and fatty acids convert myogenic cells into adipose-like cells. J Biol Chem *270:* 28183–28187.

Tontonoz, P., E. Hu, R.A. Graves, A.I. Budavari, B.M. Spiegelman (1994) mPPARγ2:Tissue-specific regulator of an adipocyte enhancer. Genes Dev *8:* 1224–1234.

Tontonoz, P., E. Hu, B.M. Spiegelman (1995) Regulation of adipocyte gene expression and differentiation by peroxisome proliferator activated receptor γ. Curr Opin Genet Dev *5:* 571–576.

Wang, N,D., M.J. Finegold, A. Bradley, C.N. Ou, S.V. Abdelsayed, M.D. Wilde, L.R. Taylor, D.R. Wilson, G.J. Darlington (1995) Impaired energy homeostasis in C/EBPα knockout mice. Science *269:* 1108–1112.

Ware, C.B., M.C. Horowitz, B.R. Renshaw, J.S. Hunt, D. Liggitt, S.A. Koblar, B.C. Gliniak, H.H. McKenna, T. Papayannopoulou, B. Thoma, et al. (1995) Targeted disruption of the low-affinity leukemia inhibitory factor receptor gene causes placental, skeletal, neural and metabolic defects and results in perinatal death. Development *121:* 1283–1299.

Weintraub, H., S.J. Tapscott, R.L. Davis, M.J. Thayer, M.A. Adam, A.B. Lassar, A.D. Miller (1989) Activation of muscle-specific genes in pigment, nerve, fat, liver, and fibroblast cell lines by forced expression of MyoD. Proc Natl Acad Sci USA *86:* 5434–5438.

Young, P., J.R.S. Arch, M. Ashwell (1984) Brown adipose tissue in the parametrial fat pad of the mouse. FEBS Lett *167:*10–14.

Cells
Tissues
Organs

Cells Tissues Organs 1999;165:181–189

Accepted after revision: September 13, 1999

Embryonic Stem Cell-Derived Haematopoiesis

Nicholas Hole

Biological Sciences, University of Durham, UK

Key Words
Differentiation, in vitro · Embryonic stem cell · Pluripotent cell types · Haematopoiesis

Abstract
Embryonal stem (ES) cells are pluripotent cell types that in chimaeric mice can give rise to all cellular lineages. Early studies showed that they also had the potential to form mature erythrocytes following in vitro differentiation. In recent years ES cells have been demonstrated to be competent both to produce all haematopoietic lineages following in vitro differentiation and long-term repopulating haematopoietic stem cells. This review discusses the approaches that have been used to produce these haematopoietic cells and the nature of the haematopoietic stem cells that can be isolated. The utility of the system to both isolate genes involved in control of haematopoiesis and to assess their function following transgenesis is demonstrated. The prospectives for this field are then discussed in the context of recent developments in human ES cells and evidence of stem cell plasticity.

Abbreviations used in this paper

AGM	aorta-gonad-mesonephros
ES	embryonal stem
HSC	haematopoietic stem cells

Introduction

Like many developmental systems, haematopoiesis is an extremly complex process. The terminally differentiated cells of the haematopoietic system comprise at least eight lineages, each with distinct functions, morphology and cell kinetics, all of which arise from a single cell type, the pluripotent haematopoietic stem cell (HSC) (fig. 1). HSC are relatively rare, numbering approximately 1 per 10^4 marrow cells [Young et al., 1996]. Although their absolute numbers may vary somewhat over the lifetime of an individual [deHaan and VanZant, 1999], this comparatively small population ultimately gives rise to approximately 10^6 cells per second in adult steady state haematopoiesis for decades on end. The balance between HSC self-renewal, quiescence and differentiation can change depending on the age of the individual [deHaan and VanZant, 1999]. These complexities are amplified in embryogenesis, where the anatomical site, the type of haematopoietic cell produced and the nature of HSC may change rapidly in the space of days [Dzierzak and Medvinsky, 1995; Medvinsky and Dzierzak, 1996].

KARGER

Fax + 41 61 306 12 34
E-Mail karger@karger.ch
www.karger.com

1422–6405/99/1654–0181$17.50/0

Accessible online at:
www.karger.com/journals/cto

Nicholas Hole, PhD
Biological Sciences, University of Durham
South Road
Durham DH1 3LE (UK)
Tel. +44 191 374 2414, Fax +44 191 374 2417, E-Mail Nicholas.Hole@durham.ac.uk

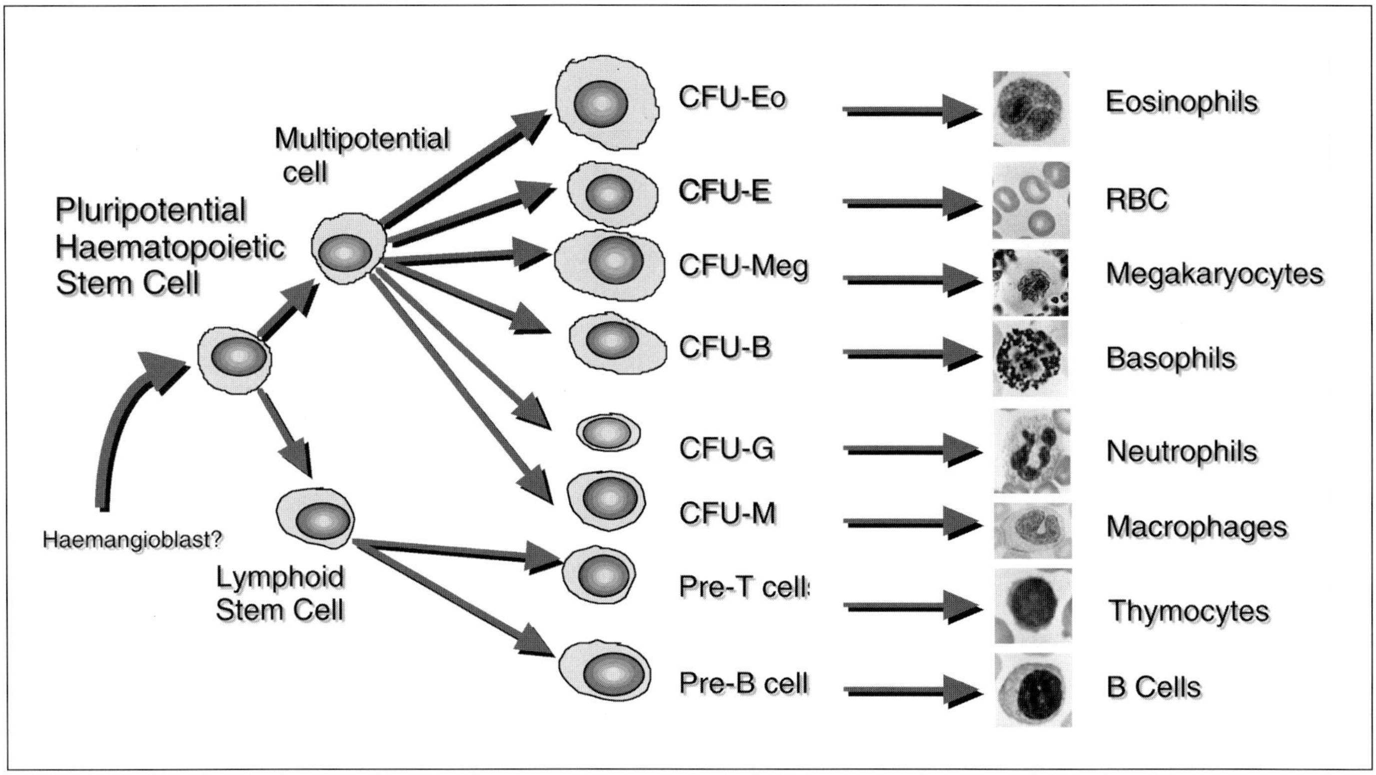

Fig. 1. A schematic of haematopoiesis. The relationship of the haemangioblast with the earliest embryonic HSC is an extremely active area of research. Note that pluripotent HSC are the only haematopoietic cells that retain both lymphoid and myeloid potential. Eo = Eosinophils; E = erythrocytes; Meg = megakaryocytes; B = basophils; G = granulocytes; M = macrophages.

The key to understanding the control of haematopoiesis is to better characterise the biology of HSC. However, direct analysis of native HSC remains a problem. Although the surface antigen phenotype of both adult and fetal HSC is becoming more closely defined [Morrison et al., 1995: Petrenko et al., 1999], reliable isolation of pure HSC populations remains difficult, in part due to the relatively small number present in haematopoietic tissues. Further, despite recent advances [Cheng et al., 1998], transgenesis of native HSC remains comparatively difficult. Directed loss of function mutagenesis (gene knockouts) of HSC is unlikely to become routine in the near future.

One approach to overcoming the problem of somatic transgenesis of HSC has been to use immortalised cell lines. Tissue culture cell lines that display limited or comparatively wide haematopoietic differentiative potential have been reported, and these have proved invaluable in examining the molecular and cellular control of haematopoiesis. Of particular note is the FDCP cell lines [Testa and Molineux, 1993], which, for example, have been used to help elucidate the cytokine control of haematopoietic commitment. However, these in vitro tissue culture-adapted systems run the risk of not reflecting the in vivo condition. In particular, they may not reflect the native HSC potential for long-term multilineage haematopoietic reconstitution in host recipients.

In addition, the analysis of both native HSC and haematopoietic cell lines in vitro does not allow us to ask basic questions of early developmental biology – from what cell type do HSC arise and how is this process controlled?

Ideally we would want a primary cell line that could produce an in vitro model of haematopoiesis demonstrating the entire differentiative potential of the haematopoietic system, produce HSC with long-term multilineage haematopoietic repopulating capacity, whose genome could be manipulated relatively easily, and which could be used to study the early commitment events that lead to the origination of HSC from more primitive precursors. As seen in other articles in this issue, and in the text below, I propose that embryonal stem (ES) cells may satis-

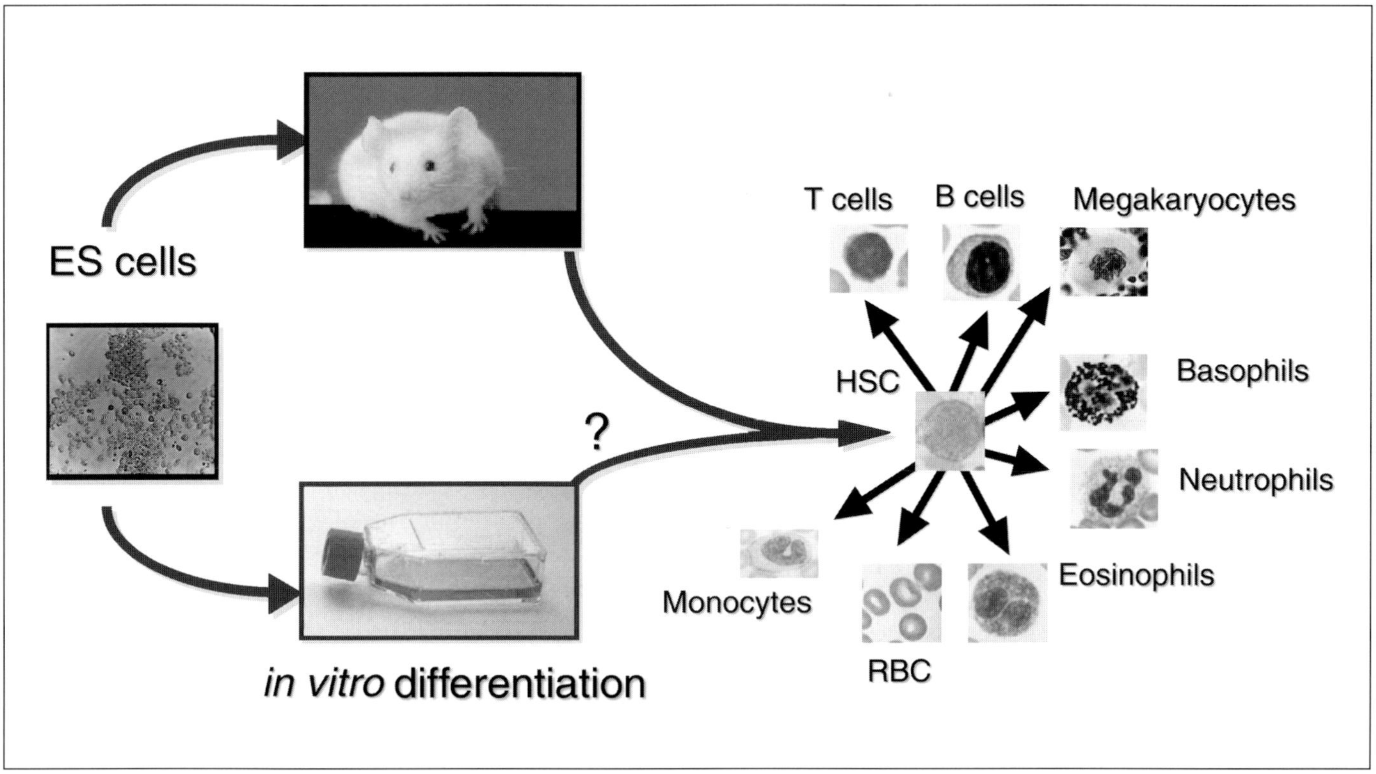

Fig. 2. ES cells are pluripotent, and as such can produce HSC in vivo following blastocyst injection and chimaera production. These can then differentiate into the different haematopoietic lineages shown. Evidence is now emerging that this potential can be exhibited following differentiation in vitro.

fy many, if not all of these criteria, and may prove to be powerful tools in the study of the molecular and cellular control of haematopoiesis (fig. 2).

The most important characteristic of an in vitro model is how closely it reflects the in vivo condition. In terms of haematopoiesis, what is the evidence that ES cells can produce the entire repertoire of the haematopoietic system in vitro, and can HSC be demonstrated?

Most, if Not All, Haematopoietic Lineages Can Be Produced Following in vitro Differentiation of ES Cells

Although ES cells will normally begin to differentiate following the withdrawal of LIF, it appears that the conditions of culture of these pluripotent cell types during that differentiation have a critical role to play in the nature of the cell lineages that are subsequently produced. At least three approaches have been used: ES cells can be aggregated, and allowed to differentiate in suspension culture, can be seeded in semisolid culture and allowed to differentiate in situ, and can be allowed to differentiate in the presence of accessory cell types [Hole and Smith, 1994].

Suspension culture was used in some of the earliest reports of in vitro differentiation of ES cells. Doetschman et al. [1985] reported the formation of cystic embryoid bodies from ES cells following the withdrawal of LIF and growth in suspension culture. These bodies contained blood islands, reminiscent of yolk sac haematopoiesis, which were made up of erythrocytes and macrophages. Differentiation in semisolid medium has been successfully used by several groups to demonstrate the production of neutrophil, mast cell, macrophage and erythroid lineages [Wiles and Keller, 1991; Keller et al., 1993a; Lieschke and Dunn, 1995]. Use of accessory cell lines, such as OP9 has been used by workers who have subsequently demonstrated the presence of erythroid, myeloid and lymphoid lineages [Nakano et al., 1994] the latter including natural killer cell types [Nakayama et al., 1998].

Although this brief summary indicates that ES cells can indeed realise the potential to form most if not all haematopoietic lineages during differentiation in vitro, it is not so clear as to whether they can do so autonomously. Several groups reported the requirement for additional haematopoietic growth factor [Wiles and Keller, 1991]. The work of Nakano and others [Nakano et al., 1994] suggests that the use of the macrophage-colony-stimulating factor-deficient cell line OP9 [derived from op/op mice; Yoshida et al., 1990] is critical to facilitating comprehensive haematopoietic differentiation. The need for stromal cells was also indicated by the workers using the RP010 stromal cell line; in this case, exogenous growth factors were also used [Palacios et al., 1995]. In contrast, other groups reported that commitment to myeloid, erythroid or lymphoid lineages appears to not require exogenous cell lines or growth factors [Hole et al., 1996a]. What is clear from these studies is the need for ES cells to form a multicellular structure, be it embryoid body, or interaction with stromal layer prior to successful haematopoietic differentiation; this may suggest that a microenvironment needs to be developed.

Apparent differences in outcome of haematopoietic differentiation may be due to several different approaches by these groups. Some workers have used exogenous cytokines which may amplify otherwise low levels of specific lineage commitment. Indeed, it is clear that the differentiating ES cells themselves contain transcripts for a wide range of haematopoietic cytokines [Hole et al., 1996a, b] and factors [Keller et al., 1993b] which can influence the commitment process. In addition, one of the powerful advantages of ES cells as tools for studying haematopoiesis is that they are primary cell lines. Although ES cells have a practical definition – the ability to form germ-line-competent animals following blastocyst injection – it is perhaps not surprising that different ES cell lines may exhibit different properties during in vitro differentiation. Finally, nearly all reports that have studied haematopoietic differentiation of ES cells in vitro have relied on fetal calf serum as a growth supplement. The work of Wiles and his coworkers [Johansson and Wiles, 1995] has suggested that different batches of fetal calf serum can have dramatic differences in haematopoietic differentiation; the characterisation of a defined medium should go some way to clarify this process.

It is thus clear that under appropriate in vitro culture conditions, totipotential ES cells can produce mature haematopoietic cell types. However, in order to confirm that this is indeed an appropriate model for studying the process (rather than the outcome) of haematopoiesis, we need to examine whether multipotent haematopoietic precursors can be detected, critically, whether HSC can be found.

ES-Derived Haematopoietic Stem Cells Are Capable of Long-Term Multilineage Reconstitution

The cellular phenotype of haematopoietic stem cells is becoming ever more closely characterised, be it by surface antigen phenotype [Morrison et al., 1995a, b] or dye efflux [Goodell et al., 1997], but the most effective definition is one of function – as a long-term multilineage haematopoietic repopulating cell type. As other groups have reported, other definitions of what a haematopoietic stem cell can be are valid [Morrison et al., 1997], but it is clear that the most primitive HSC are found within this long-term repopulating population.

There is considerable evidence that lymphoid progenitors can be produced and isolated following ES cell differentiation in vitro. Adoptive transfer into mice whose lymphoid compartment is compromised by genetic lesion has resulted in ES cell-derived lymphoid repopulation over both the long and short term [Potocnik et al., 1997]. Early reports suggested that repopulating ability of ES cell-derived haematopoietic progenitors may be restricted to the lymphoid system [Muller and Dzierzak, 1993]; however, further studies have shown ES cell-derived cells can demonstrate long-term multilineage haematopoietic repopulating potential [Palacios, et al., 1995; Hole et al., 1996a]. In the latter case, our own studies have shown that the proportion of peripheral blood leukocytes derived from ES cells, although multilineage and present over many months, was surprisingly low. The question arises as to whether these data can indicate what stage of embryonic haematopoiesis ES cell differentiation in vitro may model. In order to best address this, we need to briefly review current understanding of embryonic haematopoiesis.

Embryonic haematopoiesis is first detectable in the day-7 yolk sac. Subsequently, the aorta-gonad-mesonephros (AGM) region contains haematopoietic progenitors by day 10–12. Fetal liver is the major site of haematopoietic activity from day 12 to birth [Dzierzak and Medvinsky, 1995]. An area of intense study in the past few years has been the question as to whether haematopoietic progenitors originate de novo within these sites, or reflect the migration of the progenitors within the developing embryo. Exemplary work from Medvinsky and his coworkers has demonstrated that the AGM region is almost certainly the site from which all subsequent haematopoietic progenitors ultimately arise [Medvinsky and Dzierzak, 1996; Medvinsky et al., 1996], but what is the relationship, if any, between these intra-embryonic progenitors and yolk sac haematopoietic progenitors? AGM-derived HSC are clearly capable of long-term multilineage repopulation of irradiated recipients, but early yolk sac HSC appeared capable of only repopulating erythroid and lymphoid compartments; only by day 11 (when HSC are demonstrably present in the AGM region) can long-term repopulating HSC be detected [Muller et al., 1994]. This has led to the concept of 'primitive' HSC, capable of only limited repopulation. The initial observations of ES cell-derived haematopoietic repopulation suggested that ES cell-derived HSC were also capable of only limited lineage repopulation [Muller and Dzierzak, 1993], suggesting that the model may not be best suited for studying long-term multilineage HSC. However, all of the ES cell repopulation studies described above were carried out using adult mice as recipients. Recent published work has suggested that yolk sac HSC are indeed capable of long-term multilineage repopulating ability, but that this potential is best observed when adoptively transferred to newborn, not adult, animals [Yoder and Hiatt, 1997]. This predilection to repopulating neonates is not retained posttransplantation; bone marrow HSC taken from these primary recipients when adults repopulate secondary adult recipients in a manner indistinguishable from conventional HSC.

What can we conclude of HSC derived from in vitro differentiation of ES cells? It is clear from in vitro work that ES cell-derived haematopoietic progenitors are capable of producing cells of most, if not all lineages. Initial in vivo repopulation studies [Muller and Dzierzak, 1993] and surface antigen analysis [Potocnik et al., 1997] suggested that these may be recapitulating early embryonic haematopoiesis, probably producing yolk sac HSC. The recent data demonstrating that yolk sac HSC are best capable of multineage reconstitution of neonate, not adults, may go someway to explain our own observations of ES cell-derived haematopoietic reconstitution of adults, although multilineage is typically less than 5% of nucleated leukocytes were ES cell-derived [Hole et al., 1996a]. In subsequent studies, adoptive transfer of ES cell-derived HSC from primary recipients into secondary adults has resulted in dramatically increased ES cell-derived haematopoietic reconstitution (unpubl. results), perhaps reflecting similar data from yolk sac HSC transplantation experiments. It is clear that the microenvironment of HSC has a critical impact on their phenotype. In the experiments where yolk sac HSC were transferred to neonates and studied following the development to adults, the adult marrow environment appears to change their properties from yolk sac to adult-type HSC [Yoder and Hiatt, 1997]. In vitro coculture of yolk sac HSC with endothelial cell lines can also condition them to demonstrate multilineage repopulating ability of adult mice [Lu et al., 1996]. Interestingly, one report of comparatively high level ES cell-derived haematopoietic reconstitution of adults has relied on stromal cell line coculture in the differentiation process [Palacios et al., 1995] reinforcing the evidence of the plasticity of ES cell-derived HSC.

In summary, in vitro ES cell-derived haematopoiesis can recapitulate many, if not all the haematopoietic processes leading to the development of mature haematopoietic cell types. Further, it appears to be a useful in vitro model of very early haematopoietic commitment, and can produce HSC capable of long-term multilineage haematopoietic reconstitution. There are at least two areas where this model has been best used: analysis of gene function and screening for genes involved in haematopoietic commitment.

Analysis of Gene Function in Haematopoiesis

ES cells and gene knockouts have already proved to be powerful tools in dissecting the molecular control of many developmental systems [Hooper, 1994]. Exploiting the relative ease with which the genome of ES cells can be manipulated, and producing germ-line-competent transgenic mice following blastocyst injection, null allele mutations of specific genes can be studied [reviewed elsewhere in this issue; see also Ramirez-Solis et al., 1993]. However, in the case of several genes that are implicated in early haematopoiesis, homozygote null allele mice die early in embryogenesis; an example is Rb and vav [Clarke et al., 1992]. The issue of whether these genes are involved in later haematopoiesis is moot. However, one advantage of studying the haematopoietic system is that if embryogenesis can progress to the stage where HSC can be reliably isolated, then these can be transplanted into conditioned recipients. Mice with null alleles for DNA ligase 1 die in utero, but transplantation of fetal liver HSC from these null-allele embryos into non-transgenic conditioned adult recipients allowed the study of the role of this gene in haematopoietic development [Bently et al., 1996]. In order to produce null-allele mice, ES cells are normally injected into host blastocysts. The resulting chimaeric animal can have a variable level of ES cell-derived cellular contribution; only by subsequent breeding and backcrossing can the phenotype of the mutation be analysed fully. However, by using a 'sandwich' of ES cells and tetraploid embryonic cells, mice can be produced with very high ES cell chimaerism – approaching 100% [Nagy et al., 1990]. Although these mice can be live born, viability is often quite low [Nagy et al., 1993]. However, the fetal liver HSC from these wholly ES cell-derived mice are transplantable and can be used for further study following transplantation into conditioned recipients [Forrester et al., 1991].

Importantly, all of the above approaches require successful embryogenesis, at least to the stage where HSC are formed. However, this is only one aspect of the complex process of haematopoiesis; a critical part of the process is the stromal microenvironment in which it develops [Whetton and Spooncer, 1998] – and this cannot necessarily be transplanted so easily. An in vitro approach can overcome some of these problems. Mutations can be rendered homozygous in ES cells by double gene targeting events of high dose drug selection [Watanabe et al., 1995]. Using some of the strategies outlined earlier to produce haematopoietic cells, the role of this null mutation in the phenotype of haematopoiesis can be studied. For example, GATA factors are implicated in the control of haematopoiesis, but mutation of the x-linked GATA1 gene of homozygous mutation of GATA2 results in early embryonic death [Tsai et al., 1994; Pevny et al., 1995]. However, two-stage in vitro differentiation of mutated ES cells can reveal the role of these transcription factors in haematopoietic differentiation; in vitro erythroid differentiation of GATA1-null ES cells is blocked early in erythropoiesis, although other lineages can be produced [Tsai and Orkin 1997]. In contrast, GATA2-null ES cells can produce erythroid or some myeloid lineages, but may play an important role in controlling the numbers of multipotent progenitor cells and in the differentiation of mast cells. The molecular biology of very early commitment is extremely amenable to study by in vitro haematopoietic differentiation of ES cells. Mice lacking the transcription factor scl die very early in development, with no apparent yolk sac erythropoietic activity [Robb et al., 1996]. In vitro differentiation of scl-null ES cells demonstrated that there was a complete absence of detectable haematopoietic gene expression [Elefanty et al., 1997]. Mutations in genes that affect other lineages can also be examined for their effects in early haematopoiesis; mutations in GATA4 result in very poor haematopoietic differentiation of ES cells in vitro; however, this appears to be a consequence of a primary defect in the inability to form visceral yolk sac endoderm in the embryoid bodies rather than a cell-autonomous defect in HSC themselves [Bielinska et al., 1996].

Overexpression of a candidate gene in differentiating ES cells can also throw light on its function in haematopoiesis. Using an expression vector that drove constitutive expression, the zinc finger gene MZF-1 has been demonstrated to play an important role in regulating early haematopoietic development [Perrotti et al., 1995]. Using a similar approach, the LIM-homeobox gene LH2 appears to play a role in the control of early haematopoiesis [Pintodo et al., 1998]. Alternatively, expression of antisense transcripts can reduce gene expression levels to the point where some indication of the gene's function can be elucidated. As mentioned earlier, mice with null alleles for vav proto-oncogene die early in embryogenesis; the role of the gene in haematopoiesis is a good candidate for study by in vitro differentiation of ES cells. Expression of vav antisense transcripts in ES cells dramatically reduced the haematopoietic potential during in vitro differentiation [Wulf et al., 1993]. However, this example is also a salutary reminder of the problems of extrapolating antisense data; ES cells that have null alleles for vav appear to undergo haematopoietic differentiation in a manner practically indistinguishable from their wild-type counterparts [Zmuidzinas et al., 1995].

One of the best strategies to confirm that a phenotype present during in vitro differentiation of null-allele ES cells is due to an absence of the gene under study and not a consequence of the genome manipulation per se is to attempt to rescue the phenotype by transgenesis with the gene under study. CD34 is a sialomucin expressed on the surface of haematopoietic stem cells, but whose biology is only poorly understood. Differentiation of ES cells lacking a gene for CD34 demonstrated delays in erythroid and myeloid commitment; these were completely reversed by transfection with constructs that express wild-type CD34 [Cheng et al., 1996].

It is clear that in vitro differentiation of ES cells can be used productively to study haematopoietic function of candidate genes. However, using this approach the gene needs to be identified beforehand. Could in vitro differentiation of ES cells itself be used as a tool to identify candidate genes, as well as examine their function?

Identification of Genes Involved in Haematopoietic Differentiation

As discussed earlier, long-term repopulating HSC can be identified following differentiation of ES cells in vitro. If we were able to characterise the time course of this differentiation, then we would be in a strong position to examine the differential expression of genes at the stage at which HSC were first emerging as distinct cell type. Our own data suggests that HSC are present within a comparatively brief period of differentiation; multilineage repopulating activity is present at day 4 of differentiation, but not found either at day 3 or day 5 [Hole et al., 1996a]. Expression of known haematopoietic genes reinforces the importance of this period in haematopoietic differentiation; expression is dramatically upregulated in this period [Hole et al., 1996a; Hole and Graham, 1997]. Thus if genes known to be involved in early haematopoiesis are expressed at high levels, can we use this system as a model to isolate other genes that may play a critical role in this process? Using a subtractive hybridisation approach [Wang and Brown, 1991], Hole and Graham [1997] have demonstrated that this model of in vitro differentiation is a rich source of haematopoietic genes; interestingly at least two of the novel genes identified were expressed in embryogenesis and haematopoietic cell lines in a manner consistent with early haematopoietic commitment.

An alternative approach to the identification of genes likely to be involved in early haematopoietic commitment is to use gene trapping. In this strategy, genes are mutagenised at random by the insertion of a reporter construct into the genome of ES cells, often coupled to an expression construct conferring drug resistance. Typically, the expression profile of the 'trapped' gene is then observed following production of chimaeric animals; candidate genes can then be identified by sequencing [Skarnes, 1990; Skarnes et al., 1992; Friedrich and Soriano, 1993]. The rate-limiting step in this otherwise powerful process is the production of chimaeric animals; an alternative approach is to use in vitro differentiation of ES cells as a prescreen. Using the OP9-dependent model of in vitro ES cell haematopoietic differentiation, expression trapping of haematopoietic and endothelial cells has been demonstrated [Stanford et al., 1998]. The better definition of models of haematopoietic differentiation, particularly those that can reliably give rise to HSC, will improve the utility of this approach.

Both of these approaches give a tantalising glance at the utility of haematopoietic in vitro differentiation of ES cells as a tool to identify candidate genes, and are currently being used to further dissect the commitment events surrounding early haematopoiesis. However, the power of the system will only be fully exhibited when the role of candidate genes is confirmed by transgenesis of ES cells and analysis in vitro and in vivo. What else will the near future bring?

Prospectives

Although a common functional definition, the population of HSC contains within it cells with different capacities for self-renewal, differentiation and quiescence. These properties change with the age of the HSC [Morrison et al., 1996]. It is now emerging that telomere length may reflect these changes in HSC [Vaziri, et al., 1994; Morrison et al., 1996]. One area of very active research in HSC biology is to more closely examine this relationship [Morrison et al., 1997]. HSC generated from ES cells following differentiation in vitro, coupled with a transgenic approach, which may be comparatively difficult with conventional HSC, may prove to be particularly useful. The ability to culture ES cells for extended periods prior to haematopoietic differentiation may offer an alternative approach to the role of a mitotic clock in HSC biology.

It is likely that different levels of transgenesis of ES cells will allow new approaches to studying haematopoiesis. ES cell lines have already been produced that contain freely segregating transchromosomal elements from the human genome [Hernandez et al., 1999]. Artificial minichromosomes have also been introduced into ES cells and maintained for extended periods in culture [Shen et al., 1997]. The likelihood is that these approaches will enable more complex genetic models to be created. In particular, better mouse models of human haematopoietic disease could be produced.

As the work of Yoder and Hiatt [1997] has indicated, the cell biology of yolk sac (primitive) and adult (definitive) HSC may not be as clear as had once been thought. Work using knockouts of transcription factors such as PU.1 does suggest that primitive and definitive haematopoiesis may demonstrate differential sensitivity to the ablation of specific haematopoietic transcription factors [Fisher and Scott, 1998]. In vitro differentiation of ES cells offers one approach to dissecting the cell and molecular control of primitive and definitive haematopoiesis. In particular, it may allow the analysis of the commitment events surrounding the differentiation of haemangioblast cell types [Choi et al., 1998]. Understanding this process may have important consequences for the eventual utility of ES cells in the generation of therapeutics.

Recent work has suggested that stem cells may be remarkably plastic in their commitment to specific lineages. The remarkable work of Vescovi and colleagues [Bjornson et al., 1999] has suggested that neuronal stem cells may have the potential to exhibit properties consistent with that of haematopoietic stem cells. This challenges current models of commitment and differentiation, and suggests that specificity of a stem cell lineage may be modulated by other factors, such as the microenvironment. It is likely that the totipotentiality of ES cells will be exploited to further explore the relationship between commitment and differentiation in stem cells.

Recent reports of the isolation of human ES cell types [Thomson et al., 1998] have given these studies particular relevance. Such cell types could be used as a potent source of human HSC that could be used as adjuncts to conventional bone marrow transplantation; in particular, the potential ability to manipulate the genome of human ES cells may allow the production of universal donor HSC [Gearhart, 1998]. Human ES cells may also prove to be useful for studying early haematopoiesis. One of the strengths of murine ES cells in the study of HSC biology is the ability to test the potency of HSC produced following in vitro differentiation by adoptive transplantation into recipient animals. However, recent developments have suggested that an animal model of human haematopoietic repopulation could be useful in the study of conventional human HSC [Bhatia et al., 1997; Dick et al., 1997]; it is likely that this may be suitable for the study of potential HSC produced from human ES cells.

Acknowledgement

This work was supported in part by grants from the Wellcome Trust.

References

Bently, D.J., J. Selfridge, et al. (1996) DNA ligase I is required for foetal liver erythropoiesis but is not essential for mammalian cell viability. Nat Genet *13:* 498–491.

Bhatia, M., J.C.Y. Wang, et al. (1997) Purification of primitive human hematopoietic cells capable of repopulating immune-deficient mice. Proc Natl Acad Sci USA *94:* 5320–5325.

Bielinska, M., N. Narita, et al. (1996) Erythropoiesis and vasculogenesis in embryoid bodies lacking visceral yolk sac endoderm. Blood *88:* 3720–3730.

Bjornson, C.R.R., R.L. Rietze, et al. (1999) Turning brain into blood: A hematopoietic fate adopted by adult neural stem cells in vivo. Science *283:* 534–537.

Cheng, J., S. Baumhueter, et al. (1996) Hematopoietic defects in mice lacking the sialomucin CD34. Blood *87:* 479–490.

Cheng, L., C. Du, et al. (1998) Sustained gene expression in retrovirally transduced, engrafting human hematopoietic stem cells and their lympho-myeloid progeny. Blood *92:* 83–92.

Choi, K., M. Kennedy, et al. (1998) A common precursor for hematopoietic and endothelial cells. Development *125:* 725–732.

Clarke, A.R., E.R. Maandag, et al. (1992) Requirement for a functional RB-1 gene in murine development. Nature *359:* 328–330.

deHaan, G., G. VanZant (1999) Dynamic changes in mouse hematopoietic stem cell numbers during aging. Blood *93:* 3294–3301.

Dick, J.E., M. Bhatia, et al. (1997) Assay of human stem cells by repopulation of NOD/SCID mice. Stem Cells *15(suppl 1):* 199–203, 204–207.

Doetschman, T.C., H. Eistetter, M. Katz, W. Schmidt, R. Kemler (1985) The in vitro development of blastocyst-derived embryonic stem cell lines: Formation of visceral yolk sac, blood islands and myocardium. J Embryol Exp Morphol *87:* 27–45.

Dzierzak, E., A. Medvinsky (1995) Mouse embryonic haematopoiesis. Trends Genet *11:* 359–366.

Elefanty, A.G., L. Robb, et al. (1997) Hematopoietic-specific genes are not induced during in vitro differentiation of scl-null embryonic stem cells. Blood *90:* 1435–1447.

Fisher, R.C., E.W. Scott (1998) Role of PU.1 in hematopoiesis. Stem Cells *16:* 25–37.

Forrester, L.M., A. Bernstein, et al. (1991) Long-term reconstitution of the mouse hematopoietic system by embryonic stem cell-derived fetal liver. Proc Natl Acad Sci USA *89:* 7514–7517.

Friedrich, G., P. Soriano (1993) Insertional mutagenesis by retroviruses and promotor traps in embryonic stem cells. Methods Enzymol *225:* 681–701.

Gearhart, J. (1998) Cell biology – New potential for human embryonic stem cells. Science *282:* 1061–1062.

Goodell, M.A., M. Rosenzweig, et al. (1997) Dye efflux studies suggest that hematopoietic stem cells expressing low or undetectable levels of CD34 antigen exist in multiple species. Nat Med *3:* 1337–1345.

Hernandez, D., P.J. Mee, et al. (1999) Transchromosomal mouse embryonic stem cell lines and chimeric mice that contain freely segregating segments of human chromosome 21. Hum Mol Genet *8:* 923–933.

Hole, N., G. Graham (1997) Use of embryonal stem cells in molecular hematopoiesis. Ballières Clin Hematol *3:* 467–483.

Hole, N., G.J. Graham, et al. (1996a) A limited temporal window for the derivation of multilineage repopulating hematopoietic progenitors during embryonal stem cell differentiation in vitro. Blood *90:* 1266–1276.

Hole, N., R. Leung, et al. (1996b) Hematopoietic differentiation of embryonal stem cells in vitro; in Zander, A., W. Ostertag, B.V. Afanasiev, F. Grosveld (eds): Gene Technology. Berlin, Springer, pp 3–10.

Hole, N., A.G. Smith (1994) Embryonic stem cells and hematopoiesis; in Freshney, R.J., I.B. Pragnell, M.G. Freshney (eds): Culture of Hematopoietic Cells. New York, Wiley, pp 235–249.

Hooper, M.L. (1994) Embryonal Stem Cells. Amsterdam, Harwood Academic.

Johansson, B.M., M.V. Wiles (1995) Evidence for involvement of activin A and bone morphogenetic protein 4 in mammalian mesoderm and hematopoietic development. Mol Cell Biol *15:* 141–151.

Keller, G., M. Kennedy, et al. (1993a) Hematopoietic commitment during embryonic stem cell differentiation in culture. Mol Cell Biol *13:* 473–486.

Keller, G., M. Kennedy, et al. (1993b) Hematopoietic commitment during embryonic stem cell differentiation in culture. Mol Cell Biol *13:* 473–486.

Lieschke, G.J., A.R. Dunn (1995) Development of functional macrophages from embryonal stem cells in vitro. Exp Hematol *23:* 328–334.

Lu, L.S., S.J. Wang, et al. (1996) In vitro and in vivo differentiation into B cells. T cells, and myeloid cells of primitive yolk sac hematopoietic precursor cells expanded >100-fold by coculture with a clonal yolk sac endothelial cell line. Proc Natl Acad Sci USA *93:* 14782–14787.

Medvinsky, A., E. Dzierzak (1996) Definitive hematopoiesis is autonomously initiated by the AGM region. Cell *86:* 897–906.

Medvinsky, A.L., O.L. Gan, et al. (1996) Development of day-8 colony forming units-spleen hematopoietic progenitors during early murine embryogenesis: Spatial and temporal mapping. Blood *87:* 557–566.

Morrison, S.J., H.D. Hemmati, et al. (1995a) The purification and characterisation of fetal liver hematopoietic stem cells. Proc Natl Acad Sci USA *92:* 10302–10306.

Morrison, S.J., K.R. Prowse, et al. (1996) Telomerase activity in hematopoietic cells is associated with self-renewal potential. Immunity *5:* 207–216.

Morrison, S.J., N. Uchida, et al. (1995b) The biology of hematopoietic stem cells. Annu Rev Cell Dev Biol *11:* 35–71.

Morrison, S.J., A.M. Wandycz, et al. (1996) The aging of hematopoietic stem cells. Nat Med *2:* 1011–1016.

Morrison, S.J., D.E. Wright, et al. (1997) Hematopoietic stem cells: Challenges to expectations. Curr Opin Immunol *9:* 216–221.

Muller, A.M., E.A. Dzierzak (1993) ES cells have only a limited lymphopoietic potential after adoptive transfer into mouse recipients. Development *118:* 1343–1351.

Muller, A.M., A. Medvinsky, et al. (1994) Development of hematopoietic stem-cell activity in the mouse embryo. Immunity *1:* 291–301.

Nagy, A., E. Gocza, et al. (1990) Embryonic stem cells alone are able to support fetal development in the mouse. Development *110:* 815–821.

Nagy, A., J. Rossant, et al. (1993) Derivation of completely cell-culture derived mice from early passage embryonic stem cells. Proc Natl Acad Sci USA *90:* 8424–8428.

Nakano, T., H. Kodama, et al. (1994) Generation of lymphohematopoietic cells from embryonic cells in culture. Science *265:* 1098–1101.

Nakayama, N., I. Fang, et al. (1998) Natural killer and B-lymphoid potential in CD34+ cells derived from embryonic stem cells differentiated in the presence of vascular endothelial growth factor. Blood *91:* 2283–2295.

Palacios, R., E. Golunski et al. (1995) In vitro generation of hematopoietic stem cells from an embryonic stem cell line. Proc Natl Acad Sci USA *92:* 7530–7534.

Perrotti, D., P. Melotti, et al. (1995) Overexpression of the zinc finger protein MZF1 inhibitis hematopoietic development from embryonic stem cells: Correlation with negative regulation of CD34 and c-myb promoter activity. Mol Cell Biol *15:* 6075–6087.

Petrenko, O., A. Beavis, et al. (1999) The molecular characterization of the fetal stem cell marker AA4. Immunity *10:* 691–700.

Pevny, L., L. Chyuan-Sheng, et al. (1995) Development of hematopoietic cells lacking transcription factor GATA-1. Development *121:* 163–172.

Pintodo, O.P., A. Kolterud, et al. (1998) Expression of the LIM-homeobox gene LH2 generates immortalized steel factor-dependent multipotent hematopoietic precursors. EMBO J *17:* 5744–5756.

Potocnik, A.J., G. Nerz, et al. (1997) Reconstitution of B cell subsets in Rag-deficient mice by transplantation of in vitro differentiated embryonic stem cells. Immunol Lett *57:* 131–137.

Ramirez-Solis, R., A.C. Davis, et al. (1993) Gene targeting and embryonal stem cells. Methods Enzymol *225:* 855–878.

Robb, L., N.J. Elwood, et al. (1996) The scl gene product is required for the generation of all hematopoietic lineages in the adult mouse. EMBO J *15:* 4123–4129.

Shen, M.H., J. Yang, et al. (1997) Human minichromosomes in mouse embryonal stem cells. Hum Mol Genet *6:* 1375–1382.

Skarnes, W. (1990) Entrapment vectors – A new tool for mammalian genetics. Biotechnology *8:* 827–831.

Skarnes, W., A. Auerbach, et al. (1992) A gene trap approach in mouse embryonic stem cells – The lac-Z reporter is activated by splicing, reflects endogenous gene-expression and is mutagenic in mice. Genes Dev *6:* 903–918.

Stanford, W.L., G. Caruana, et al. (1998) Expression trapping: Identification of novel genes expressed in hematopoietic and endothelial lineages by gene trapping in ES cells. Blood *92:* 4622–4631.

Testa, N.G., G. Molineux (1993) Haemopoiesis: A Practical Approach. Oxford, IRL Press.

Thomson, J.A., J. ItskovitzEldor, et al. (1998) Embryonic stem cell lines derived from human blastocysts. Science *282:* 1145–1147.

Tsai, F.Y., G. Keller, et al. (1994) An early haematopoietic defect in mice lacking the transcription factor GATA-2. Nature *371:* 221–226.

Tsai, F.Y., S.H. Orkin (1997) Transcription factor GATA-2 is required for proliferation/survival of early hematopoietic cells and mast cell formation, but not for erythroid and myeloid terminal differentiation. Blood *89:* 3636–3643.

Vaziri, H., W. Dragowska, et al. (1994) Evidence for a mitotic clock in human hematopoietic stem-cells – Loss of telomeric DNA with age. Proc Natl Acad Sci USA *91:* 9857–9860.

Wang, Z., D.D. Brown (1991) A gene expression screen. Proc Natl Acad Sci USA *88:* 11505–11509.

Watanabe, S., N. Kai, et al. (1995) Stable production of mutant mice from double gene converted ES cells with puromycin and neomycin. Biochem Biophys Res Commun *213:* 130–137.

Whetton, A.D., E. Spooncer (1998) Role of cytokines and extracellular matrix in the regulation of haemopoietic stem cells. Curr Opin Cell Biol *10:* 721–726.

Wiles, M.V., G. Keller (1991) Multiple hematopoietic lineages develop from embryonal stem (ES) cells in culture. Development *111:* 259–267.

Wulf, G.M., C.N. Adra, et al. (1993) Inhibition of hematopoietic development from embryonic stem cells by antisense vav RNA. EMBO J *12:* 5065–5074.

Yoder, M.C., K. Hiatt (1997) Engraftment of embryonic hematopoietic cells in conditioned newborn recipients. Blood *89:* 2176–2183.

Yoshida, H., S.-I. Hayashi, et al. (1990) The murine mutation osteopetrosis is in the coding region of the macrophage colony stimulating factor gene. Nature *345:* 442–444.

Young, J.C., D. DiGiusto, et al. (1996) In vitro characterisation of fetal hematopoietic stem cells: Range and kinetics of cell production from individual stem cells. Biotechnol Bioeng *50:* 465–478.

Zmuidzinas, A., K.D. Fischer et al. (1995) The vav proto-oncogene is required early in embryogenesis but not for hematopoietic development in vitro. EMBO J *14:* 1–11.

Cells
Tissues
Organs

Cells Tissues Organs 1999;165:190–202

Accepted after revision: June 7, 1999

Induction of Cellular Differentiation by Retinoic Acid in vitro

Jürgen Rohwedel[a] Kaomei Guan[b] Anna M. Wobus[b]

[a]Department of Medical Molecular Biology, University of Lübeck, and [b]In Vitro Differentiation Group, IPK Gatersleben, Germany

Key Words

Embryonic stem cells · Differentiation · Retinoic acid · Cardiogenesis

Abstract

Cellular differentiation by the vitamin A derivative retinoic acid (RA) has been studied with undifferentiated pluripotent embryonic carcinoma (EC) and embryonic stem (ES) cells in vitro. Both cellular systems are suitable to study differentiation of various cell types, because they recapitulate early stages of mouse embryogenesis. In vivo, RA was identified as a morphogenic and teratogenic compound and furthermore as a signalling molecule influencing gene expression in a complex manner via a family of RA receptors. Here, we summarize in vitro studies with ES and EC cells in comparison to in vivo studies that have contributed to our understanding how RA influences differentiation and regulates gene expression. We demonstrate that modulation of ES cell differentiation in vitro by RA depends on the concentration and developmental stage of application which is comparable to its stage-dependent influence on embryonic development in vivo.

Abbreviations used in this paper

BMP	bone morphogenetic protein
CRABP	cellular retinoic acid binding protein
ECC	embryonic carcinoma cells
ESC	embryonic stem cells
RA	retinoic acid
RAR	retinoic acid receptor
RXR	retinoid X receptor

Introduction

A prerequisite to study cellular differentiation in vitro is the availability of model systems that closely resemble development in vivo. Mouse embryonic stem (ES) cells, derived from the inner cell mass of mouse blastocysts, are undifferentiated, pluripotent cells that have the capacity to differentiate into cell types of all three primary germ layers [Doetschman et al., 1985]. In vivo, ES cells have therefore been used for gene targeting experiments to generate genetically altered mice [Thomas and Capecchi, 1987]. In vitro, ES cells spontaneously differentiate into many cell types and were used as a model system to study differentiation of cardiac [Wobus et al., 1991; Maltsev et al., 1993, 1994; Wobus and Guan, 1998], myogenic [Rohwedel et al., 1994, 1995, 1998b], hematopoietic [Schmitt et al., 1991; Wiles, 1993], epithelial [Bagutti et al., 1996], neuronal [Bain et al., 1995; Fraichard et al., 1995; Strübing et al., 1995; Okabe et al., 1996], vascular smooth muscle [Drab et al., 1997] and adipogenic [Dani et al.,

KARGER
Fax + 41 61 306 12 34
E-Mail karger@karger.ch
www.karger.com

1422–6405/99/1654–0190$17.50/0

Accessible online at:
www.karger.com/journals/cto

Dr. Jürgen Rohwedel
Department of Medical Molecular Biology
Medical University of Lübeck
Ratzeburger Allee 160, D–23538 Lübeck (Germany)
Tel. +49 451 5004095, Fax +49 451 5003637, E-Mail rohwedel@molbio.mu-luebeck.de

1997] cells (fig. 1). In contrast, undifferentiated teratocarcinoma-derived embryonic carcinoma (EC) cells fail to differentiate into germ cells in vivo. Furthermore, EC cells have to be induced to differentiate in vitro.

During ES cell differentiation, tissue-specific genes, proteins, ion channels and receptors are differentially expressed in a pattern closely resembling the pattern observed during mouse embryogenesis [reviewed by Guan et al., 1999; see fig. 1]. This model system enables one to study modulation of differentiation by 'gain-of-function' and 'loss-of-function' approaches and the influence of exogenous factors on differentiation such as differentiation factors, growth factors, signalling molecules and extracellular matrix proteins [e.g. Simon et al., 1992; Braun and Arnold, 1994; Rohwedel et al., 1994; Wobus et al., 1994b; Rohwedel et al., 1995; Soudais et al., 1995; Fässler et al., 1996; Wobus et al., 1997; Rohwedel et al., 1998a]. With this system, we studied the influence on differentiation of one of the most important differentiation-inducing factors, the vitamin A derivative retinoic acid (RA).

Since the beginning of this century vitamin A has been known to be an important nutritional compound that particularly plays an important physiological role during the visual cycle [reviewed by Wolf, 1996]. Vitamin A deprivation during pregnancy or the early postnatal phase results in embryonic malformations [Wilson and Arkany, 1949; Wilson et al., 1953]. The finding that administration of RA could prevent these malformations suggested that RA was the biologically active form of vitamin A. RA also has a teratogenic capacity since embryonic malformations were observed when an excess of retinoids was applied [Lammer et al., 1985]. More recently, it has become clear that RA functions as a hormone with important regulatory functions during embryonic development [reviewed by Morriss-Kay and Sokolova, 1996]. Evidence exists that RA is a morphogen that forms a concentration gradient along embryonic axes and affects the anteroposterior patterning of the body axis and the limbs [Kessel and Gruss, 1991; Chen et al., 1992; Chen and Solursh, 1992; Hogan et al., 1992]. In vitro, RA induces differentiation of EC cells and ES cells into specific cell types in a time- and concentration-dependent manner [Strickland and Mahdavi, 1978; Edwards et al., 1983; Jones-Villeneuve et al., 1983; Rudnicki and McBurney, 1987; Wobus et al., 1994a, b; Blank et al., 1995; Fraichard et al., 1995; Strübing et al., 1995; Wobus et al., 1997; Dani et al., 1997; Drab et al., 1997].

Furthermore, retinoids can influence carcinogenesis. It has been shown that transformation of cells by chemical substances can be prevented by application of retinoids. In vivo, retinoids were found to inhibit tumor development in animals and humans and therefore may serve as therapeutic agents in cancer chemoprevention [reviewed by Lotan, 1996].

The diverse effects of RA on embryonic development in vivo and cellular differentiation in vitro might be explained by complex mechanisms that regulate gene expression. RA is known to influence gene expression via two families of nuclear RA receptors, the retinoic acid receptors (RARs) and the retinoid X (9-*cis* RA) receptors (RXRs). Numerous isoforms of these receptors exist that bind to response elements of RA target genes [reviewed by Chambon, 1996].

Here, we summarize the teratogenic and morphogenic effects of RA in vivo, and the effects of RA on EC and ES cell differentiation in vitro. Furthermore, we briefly discuss the differentiation-inducing and antitumorigenic effects of RA and its molecular mechanism of action.

Teratogenic Effects of RA

Application of RA during embryogenesis resulted mainly in severe defects of craniofocial structures including the central nervous system [Lammer et al., 1985; Durston et al., 1989], the limbs [Kochhar, 1973; Alles and Sulik, 1989; Eichele, 1989; Tabin, 1991] and the axial skeleton [Sive et al., 1990; Kessel and Gruss, 1991; Kessel, 1992]. From 154 human females exposed to isotretinoin during pregnancy only a minority (~ 17%) had normal infants whereas most (~ 69%) had abortions and 14% of the newborns showed malformations [Lammer et al., 1985]. Malformations were observed in the central nervous system (86%), in craniofacial structures (81%), in the heart (57%) and in the thymus (33%). In mice, the skeletal malformations were not restricted to carniofacial structures but extended to the vertebrae and especially the limbs [Kochhar, 1973]. The multiple teratogenic effects of RA were strongly dependent on the concentration and time of application. For example, treatment of mouse pregnant females with all-*trans* RA at day 10 of gestation resulted in a complex pattern of malformations including craniofacial structures and the loss of vertebrae, whereas treatment between day 12 and 14 induced primarily limb defects [Kochhar, 1973]. Since most of the affected structures are derived from the neural tube and its derivatives, such as the neural crest cells, one might speculate that these first visible axial structures are most sensitive to RA.

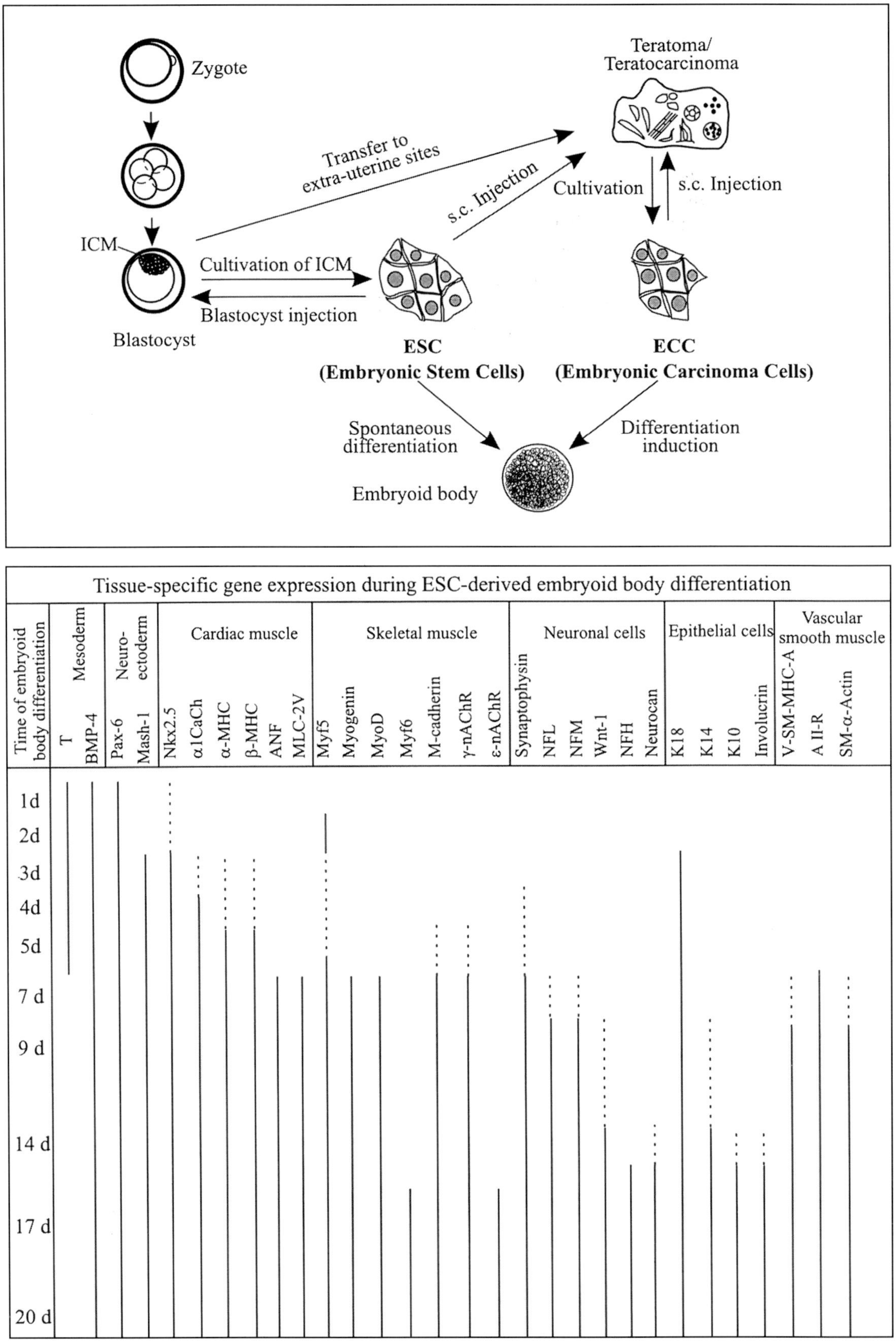
Zygote
ICM
Blastocyst
Transfer to extra-uterine sites
Cultivation of ICM
Blastocyst injection
s.c. Injection
Teratoma/ Teratocarcinoma
Cultivation
s.c. Injection
ESC
(Embryonic Stem Cells)
ECC
(Embryonic Carcinoma Cells)
Spontaneous differentiation
Differentiation induction
Embryoid body
Tissue-specific gene expression during ESC-derived embryoid body differentiation
Time of embryoid body differentiation
Mesoderm
T
BMP-4
Neuro- ectoderm
Pax-6
Mash-1
Cardiac muscle
Nkx2.5
α1CaCh
α-MHC
β-MHC
ANF
MLC-2V
Skeletal muscle
Myf5
Myogenin
MyoD
Myf6
M-cadherin
γ-nAChR
ε-nAChR
Neuronal cells
Synaptophysin
NFL
NFM
Wnt-1
NFH
Neurocan
Epithelial cells
K18
K14
K10
Involucrin
Vascular smooth muscle
V-SM-MHC-A
A II-R
SM-α-Actin
1d
2d
3d
4d
5d
7 d
9 d
14 d
17 d
20 d
1

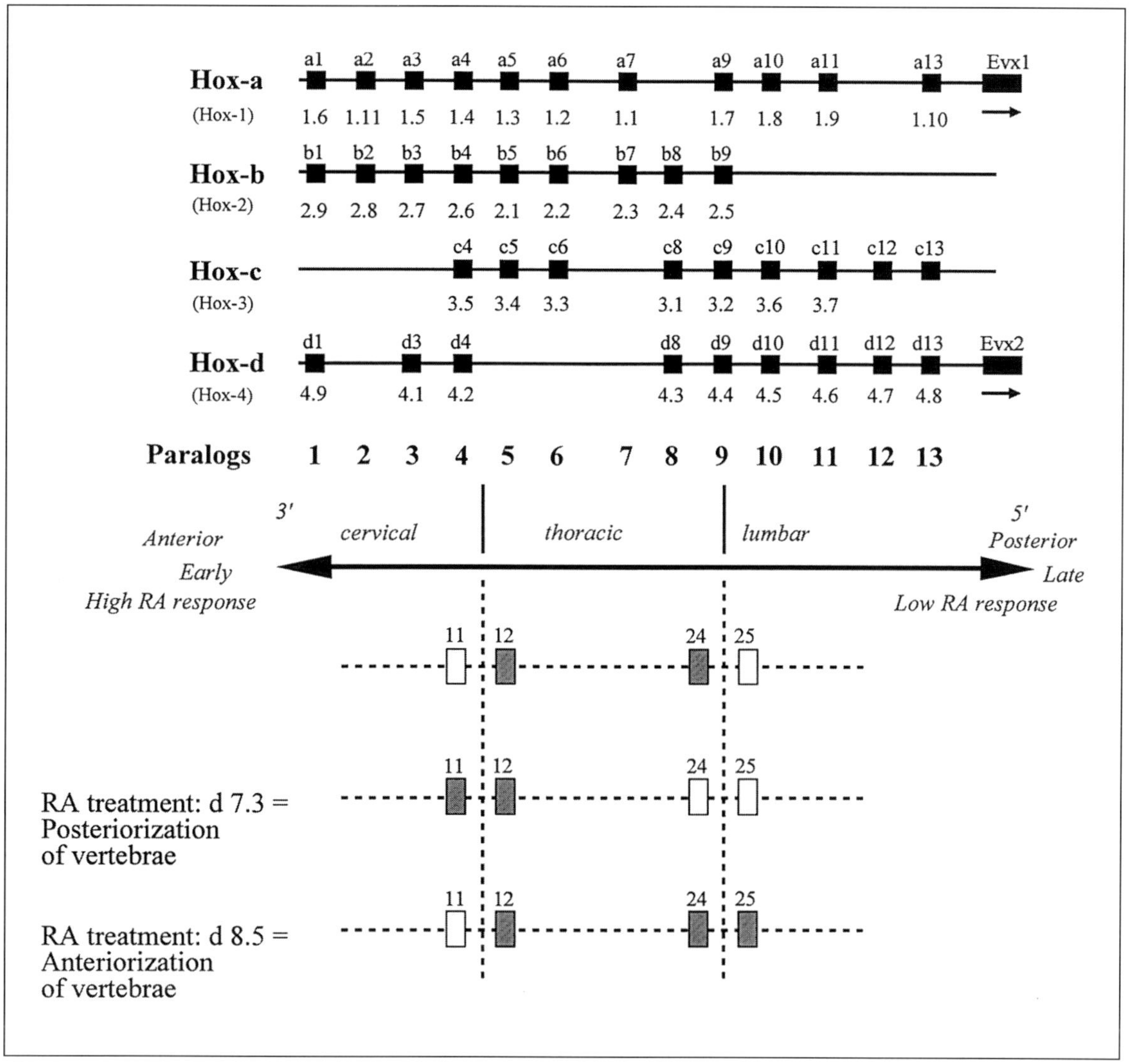

2

Fig. 1. ES cell differentiation in vitro recapitulates early embryonic development in vivo. ES cell-derived embryoid bodies differentially express genes characteristic for early mesodermal *(T, BMP-4)* and neuroectodermal development *(Pax-6, Mash-1)* and for cardiac *(Nkx2.5, α1CaCh, α-MHC, β-MHC, ANF* and *MLC-2V)*, myogenic *(Myf5, Myogenin, MyoD, Myf6, M-cadherin, γ-nAChR* and *ε-nAChR)*, neuronal *(NFL, NFM, Wnt-1, NFH, Synaptophysin, Neurocan)*, epithelial (keratins *K18, K14, K10* and *involucrin)* and vascular smooth muscle cell differentiation *(V-SM-MHC-A, AII-R, SM-α-Actin)* in a pattern closely resembling the pattern observed during mouse embryogenesis. s.c. = Subcutaneous; ICM = inner cell mass; T = Brachyury; BMP-4 = bone morphogenic protein 4; α1CaCh = α_1 subunit of the L-type calcium channel; MHC = myosin heavy chain; ANF = atrial natriuretic factor; MLC = myosin light chain; nAChR = nicotinic acetylcholine receptor; NFL = 68-kDa neurofilament protein; NFM = 160-kDa neurofilament protein; NFH = 200-kDa neurofilament protein; V-SM-MHC = vascular smooth muscle myosin heavy chain; A II-R = angiotensin II receptor; SM = smooth muscle [data from Maltsev et al., 1993, 1994; Rohwedel et al., 1994; Rose et al., 1994; Strübing et al., 1995; Bagutti et al., 1996; Drab et al., 1997; Wobus et al., 1997; Rohwedel et al., 1998a, b].

Fig. 2. RA modulates *Hox* gene expression and axis determination. *Hox* genes (filled boxes), assigned to 13 paralogous groups based on their location and sequence homology, are depicted in the upper panel in comparison to the colinearity of their spatiotemporally expression along the anteroposterior axis and their response to RA. Homeotic transformations that result either in posteriorization or anteriorization of vertebrae are caused by application of RA to pregnant female mice at day 7.3 or 8.5 of gestation, respectively, as schematically demonstrated for vertebrae 11, 12, 24 and 25 in the lower panel. The phenotypes of the last cervical and first lumbar vertebrae are presented as open boxes and the phenotypes of the first and last thoracic vertebrae are presented as filled boxes. After application of RA at day 7.3 the phenotype of vertebra 11 shifted from the last cervical (first line, open box 11) into the first thoracic vertebra (second line, filled box 11) and the phenotype of vertebra 24 from the last thoracic (first line, filled box 24) into the first lumbar vertebra (second line, open box 24). Thus, both 11 and 24 achieve a posterior phenotype. Vice versa, after application of RA at day 8.5 vertebra 25 changed its phenotype from the first lumbar (first line, open box 25) into the phenotype of the last thoracic vertebra (third line, filled box 25), a more anterior phenotype [modified from Krumlauf, 1994].

Influence of RA on Anteroposterior Pattern Formation in vivo

The most striking effect of RA was observed in the axis formation of *Xenopus* [Sieve et al., 1990] and mice [Kessel and Gruss, 1991; Kessel, 1992; Marshall et al., 1992] where it was accompanied by an altered expression of *Hox* genes (fig. 2). *Hox* genes encode transcription factors which contain a conserved DNA binding homeodomain [Scott et al., 1989]. They are arranged in clusters in the genome and are expressed in vertebrates along the dorsal axis in a coordinated manner, so that the 3′-located genes are expressed earlier and in the anterior region whereas the 5′-located genes are expressed later and in the posterior region (fig. 2). Thus, along the embryo axis different sets of *Hox* genes are active, specifying all the regions of the embryo including specific subsets of organs [Hunt and Krumlauf, 1991; Kessel and Gruss, 1991]. RA treatment in mice resulted in homeotic transformations of vertebrae depending on the concentration and the embryonic stage at the time of RA application. For example, application of RA at day 7.3 of gestation resulted in posterior transformations, whereas application at day 8.5 resulted in anterior transformations or agenesis of posterior segments (fig. 2). Posterior transformations were accompanied by an anterior shift of *Hox* gene expression and anterior transformations were accompanied by downregulation of *Hox* gene expression in the posterior part [Kessel and Gruss, 1991; Kessel, 1992; Krumlauf, 1994].

During chicken embryogenesis RA is synthesized by Hensen's node, a structure important for body axis induction and formation [Chen et al., 1992]. Grafting experiments have shown that the mouse and avian node can induce digit duplications in the avian wing similar to those produced by local administration of RA [Hogan et al., 1992]. In addition, at least in some cases, mice carrying 'loss-of-function' mutations in *Hox* genes showed transformations of skeletal elements along the anteroposterior axis [Le Mouellic et al., 1992; Jeannotte et al., 1993; Ramirez-Solis et al., 1993]. Since *Hox* gene expression can be induced by RA and, furthermore, RA response elements which bind RAR/RXR heterodimers have been detected in *Hox* genes [Yu et al., 1991; Pöpperl and Featherstone, 1993; Marshall et al., 1994], it seems possible that RA plays an important role in the regulation of *Hox* gene expression and thus in the specification of the body axis. However, it is also clear that embryonic axis formation is a complex process regulated by additional factors such as fibroblast growth factors, bone morphogenetic proteins (BMPs), sonic hedgehog and signalling molecules of the Wnt family [Sasai and De Robertis, 1997].

Concentration- and Time-Dependent Effects of RA on the Differentiation of EC and ES Cells in vitro

In accordance with the data obtained in vivo, RA influenced in vitro the expression of *Hox* genes in EC cells in a way that the 3′-located *Hox* genes are activated at lower concentrations of RA than the genes located towards the 5′ end [Breier et al., 1986; Simeone et al., 1990; Bonicelli et al., 1991]. In addition, developmentally regulated RA target genes were isolated from EC cells after RA-induced differentiation [Boudjelal et al., 1997]. Furthermore, similar to the in vivo situation, RA showed a concentration- and time-dependent effect on differentiation of EC and ES cells in vitro [Edwards and McBurney, 1983; Rudnicki and McBurney, 1987; Wobus et al., 1994b].

When RA was applied at a concentration of 10^{-7} to 10^{-8} *M* during the first 2 days of ES [Strübing et al., 1995] and EC [Jones-Villeneuve et al., 1983] cell-derived embryoid body development, neurogenesis was induced, whereas cardiogenesis was strongly inhibited (fig. 3). Recently, it has been shown that retinoid receptors are critical for neural posteriorization [Blumberg et al., 1997] and evidence exists that neuronal differentiation is controlled by anteroposterior patterning and requires posteriorizing signals [Papalopulu and Kintner, 1996].

In vivo this picture might be further complicated since evidence has been presented that neural induction and posteriorization are independent events [Kolm and Sive, 1997]. Neuronal induction might be caused by factors such as noggin, chordin or follistatin that inactivate BMPs [Piccolo et al., 1996; Kolm and Sieve, 1997]. The absence of BMPs could be sufficient to induce neuronal differentiation [Hemmati-Brivanlou and Melton, 1997].

With respect to inhibition of cardiac differentiation, posteriorization caused by RA may result in a lack of mesodermal precursors from which cardiomyocytes develop. In fact, cardiac cells develop from anterior lateral plate mesoderm. A block of cardiac differentiation was also reported for *Xenopus* embryos treated with RA at an early stage of development [Drysdale et al., 1997]. Furthermore, myocardial embryopathic defects were observed in malformed human infants after exposure to retinoids during pregnancy [Lammer et al., 1985]. These results suggested that induction of neuronal differentia-

EB stages	RA	NC	SMC	APC	CMC	VSMC
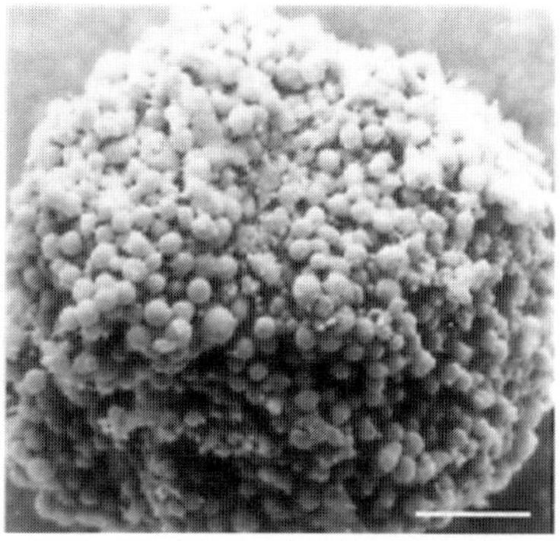	0 - 2 d	++ 10^{-7}	±	–	– –	±
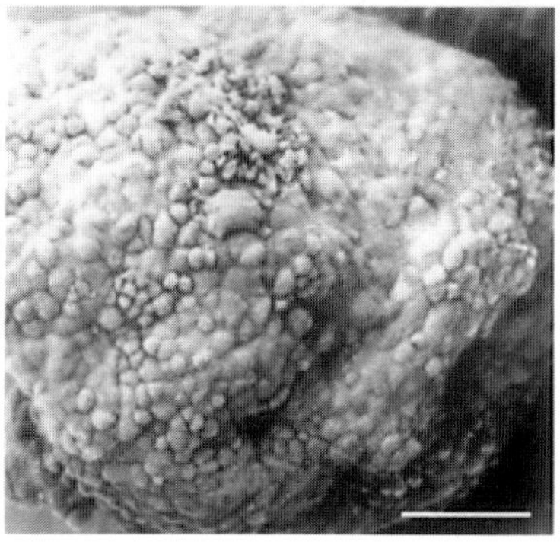	2 - 5 d	+ 10^{-8}	++ 10^{-8}	++ 10^{-8}	– –	n.d.
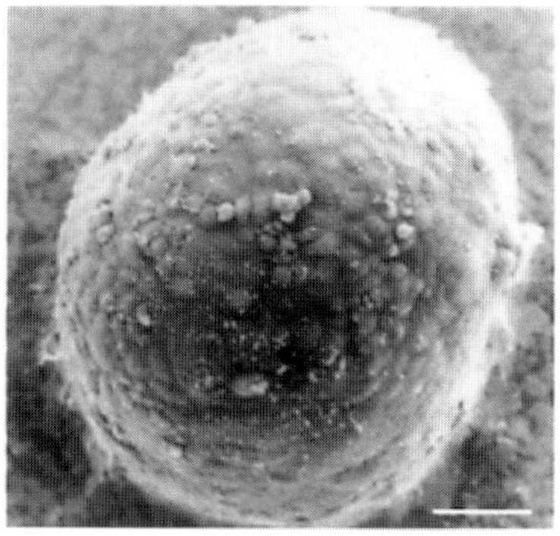	> 5 d	±	–	–	+ 10^{-8} 10^{-9}	+ 10^{-8}

Fig. 3. Stage-specific influence of RA on ES cell differentiation. ES cells were differentiated via embryoid bodies (EBs). Scanning electron micrographs of 2-, 5- and 7-day-old embryoid bodies are shown on the left. Embryoid bodies were treated with different concentrations of all-*trans* RA during defined cultivation stages: from day 0 to 2, day 2 to 5 and after day 5. The influence of RA treatment on the differentiation efficiency of neuronal cells (NC), skeletal muscle cells (SMC), adipocytes (APC), cardiomyocytes (CMC) and vascular smooth muscle cells (VSMC) in comparison to untreated cultures is shown. ± = No effect; – = inhibition; –– = strong inhibition; + = induction; ++ = strong induction. Black bars indicate induction of differentiation. Bars = 50 μm [original data: Wobus et al., 1994b, 1997; Strübing et al., 1995; Dani et al., 1997; Drab et al., 1997].

tion and inhibition of cardiogenesis may be both caused by a posteriorizing effect of RA treatment at this (early) stage of development.

Treatment of ES cell-derived embryoid bodies from day 2 to 5 with 10^{-8} *M* RA still resulted in induction of neurogenesis and inhibition of cardiogenesis. However, in addition, skeletal muscle cells [Wobus et al., 1994b] and adipocytes [Dani et al., 1997] were now induced (fig. 3). A time-dependent induction of skeletal muscle cells was also described for EC cells [Edwards and McBurney, 1983; Edwards et al., 1983; Rudnicki and McBurney, 1987]. Skeletal muscle cells and adipocytes originate from the dorsal somatic mesoderm and probably share the same mesodermal stem cell precursors [Taylor and Jones, 1979], whereas cardiomyocytes are derived from lateral plate mesoderm. Induction of skeletal muscle cell differentiation and inhibition of cardiomyocyte differentiation at the same time indicates that a developmental switch may take place at this stage. Inhibition of cardiomyocyte differentiation can be obtained by treatment of embryoid bodies with RA between day 0 and 2 and between day 2 and 5. Indeed, during these stages of embryoid body differentiation expression of mesodermal marker genes such as Brachyury and BMP-4 showed a maximum [Rohwedel et al., 1998a], suggesting that mesodermal specification takes place.

When embryoid bodies were treated with 10^{-8} *M* RA after day 5, myogenic and adipogenic differentiation was found to be inhibited, but differentiation of cardiac [Wobus et al., 1997] and of vascular smooth muscle cells [Drab et al., 1997] was induced (fig. 3). Similarly, EC cells have been induced to differentiate into vascular smooth muscle cells [Blank et al., 1995]. Induction of cardiac differentiation was even more prominent after treatment of embryoid bodies with 10^{-9} *M* RA (fig. 3). Thus, RA treatment after day 5 of embryoid body development resulted in an opposite effect compared to treatment between day 2 and 5. Similarly, in vivo, treatment of mouse embryos at day 8.5 (in comparison to day 7.3) resulted in an anteriorizing instead of a posteriorizing effect on axis formation (fig. 2).

It had already been reported more than 40 years ago that vitamin A deficiency in vivo resulted in abnormal development of the ventricular chamber of the heart [Wilson and Warkany, 1949]. In vitro, ES cell-derived cardiomyocytes undergo specialization into atrial-, ventricular-, sinusnodal- and Purkinje-like cells [reviewed by Wobus and Guan, 1998]. Therefore, we asked the question of what type of specialized cardiac cells was induced by RA. Expression of the ventricle-specific gene MLC-2V was found to be upregulated during RA-induced cardiogenic ES cell differentiation (fig. 4). Upregulation of this gene and an increase of the amount of ventricle cells [Wobus et al., 1997] indicated a specific effect of RA on cardiac development. In vivo, abnormal development of the ventricular chamber of the heart, very similar to the phenotype reported for vitamin A-deficiency [Wilson and Warkany, 1949], was observed in mice lacking the RXRα nuclear receptor [Kastner et al., 1994; Sucov et al., 1994]. Furthermore, ventricular defects with a milder phenotype compared to $RXR\alpha^{-/-}$ mice were found in RARα/RARβ double-mutant mice [Lee et al., 1997]. These results demonstrate that RA plays an important role during formation of the ventricular chamber. In addition to this major defect, mice deficient for RXRα and RARα/RARβ-double mutants [Gruber et al., 1996; Lee et al., 1997] showed abnormal development of the conotruncus, the atrioventricular canal and the aortic sac (fig. 5).

Induction of Differentiation and Anticarcinogenic Activity of RA

As mentioned above, RA is able to induce differentiation of teratocarcinoma-derived EC cell lines. Similarly, other tumor-derived or transformed cell lines were induced to differentiate by RA. For example, the neuroblastoma-derived cell line NB2a can be induced to form branching neurites [Shea et al., 1985], and the leukemia cell line HL-60 was differentiated into granulocytes by treatment with RA [Bollag, 1994]. Induction of differentiation was most efficient when retinoids were combined with cytokines such as interleukines or tumor necrosis factor-α [Lotan, 1980; Bollag, 1994]. During the last years,

Fig. 4. RA treatment resulted in enhanced development of ventricle-like cells during ES cell-derived cardiac differentiation. Enhanced activity of the reporter construct pGNA/MLC2.1 containing a ventricle-specific promoter and the *lacZ* gene (**a–d**) and enhanced expression of the endogenous *MLC-2V* gene encoding the ventricular myosin light chain-2 isoform (**e, f**) during differentiation of a stably transfected ES cell clone after treatment with 10^{-9} and 10^{-8} *M* all-*trans* RA in comparison to (untreated) control cells. Treatment of embryoid bodies (EBs) with 10^{-9} *M* all-*trans* RA between days 5 and 15 resulted in an increase of β-galactosidase-positive cells in EB outgrowths (**b**) compared to untreated control cells (**a**) at day 15. The percentage of β-galactosidase-positive EBs (**c**) and the β-galactosidase activity in relation to cellular protein content (**d**) was evalu-

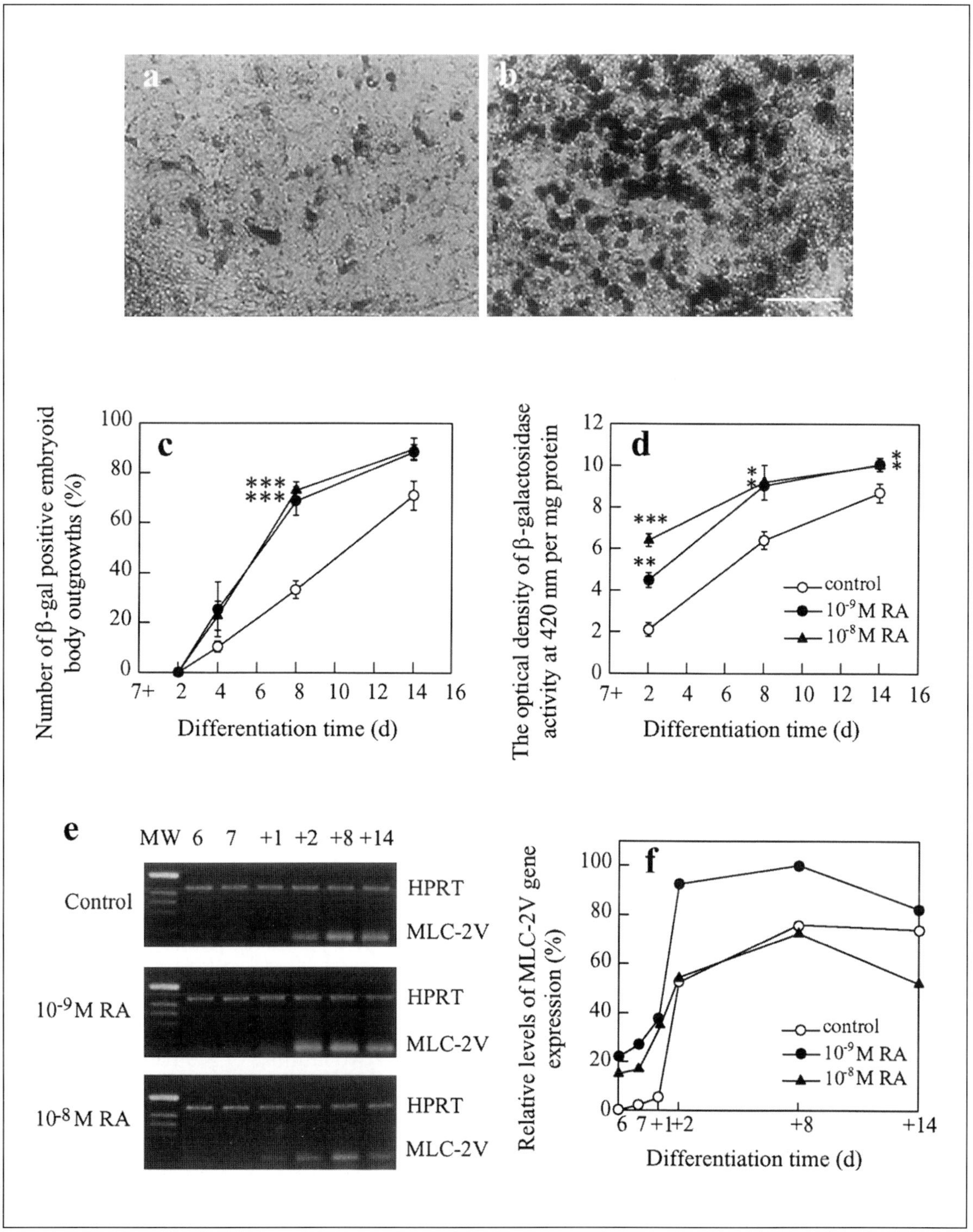

ated in control and RA-treated variants (10^{-9} and 10^{-8} *M* all-*trans* RA). Each data point represents the mean value ± standard error of the mean. Statistical significance was tested by Student's t test. Significance levels: * $p \leq 0.05$; ** $p \leq 0.01$; *** $p \leq 0.001$. The level of *MLC-2V* gene expression of 10^{-9} or 10^{-8} *M* RA-treated EBs and outgrowths of the pGNA/MLC2.1-derived ES cell clone compared to control (untreated) cells analyzed by semiquantitative RT-PCR. The ethidium bromide-stained gels of PCR products (**e**) were analyzed by computer-assisted densitometry and the data were plotted (in %) for *MLC-2V* in relation to hypoxanthine guanine phosphoribosyl-transferase *(HPRT)* gene expression, the latter used as an internal standard (**f**). Similar data were obtained in two independent experiments, and for the densitometric evaluation three independent gels were analyzed from the experiment shown in **e**. MW = Molecular weight markers. Bar (**a, b**) = 100 μm [**c–f**: original data from Wobus et al., 1997].

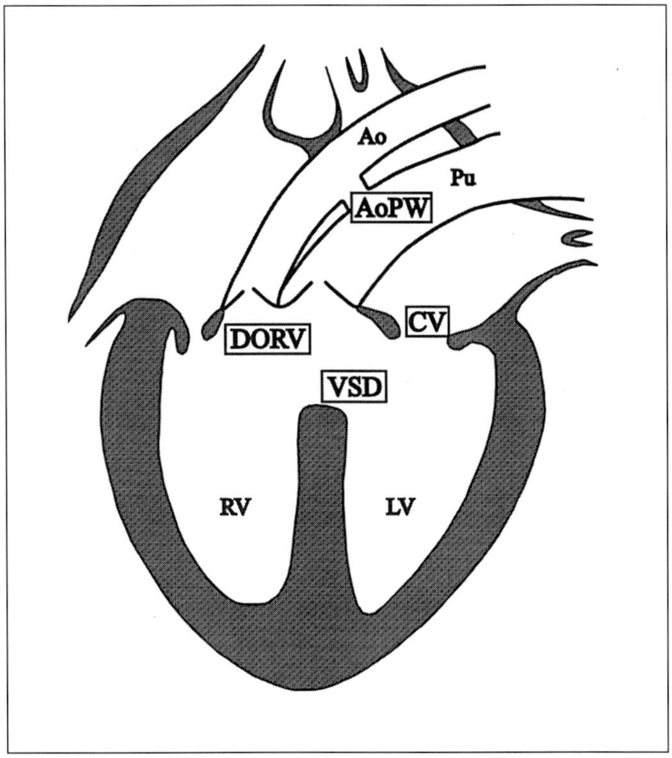

Fig. 5. Mice lacking the RXRα retinoid receptor display defects of heart development regarding mainly the ventricular chamber but also the endocardial cushions. Defects observed in hearts of mice deficient for the nuclear RA receptor RXRα and in RARα/RARβ double-mutant mice include abnormal development of (1) the ventricular chamber including ventricular septal defects (VSD), hypoplastic ventricular wall and abnormal differentiation of trabeculae and papillary muscles, (2) the aorticopulmonary septum with an aorticopulmonary window (AoPW) or persistent truncus arteriosus, (3) the conotruncus with absence of conotruncal ridges or double outlet right ventricle (DORV) and (4) the atrioventricular cushions with a common atrioventricular canal or cleft valves (CV). Ao = Aorta; Pu = pulmonalis; RV = right ventricle; LV = left ventricle [modified from Gruber et al., 1996].

all-*trans* RA has been established as a successful therapeutic agent in the treatment of acute promyelocytic leukemia in humans [Huang et al., 1988; Degos et al., 1990]. Furthermore, retinoids have been shown to prevent cancer development in head/neck and lung cancer patients [Lotan, 1996]. Inhibition of proliferation and induction of differentiation of tumor cells by RA may both rely on the ability of RA to modulate gene expression as discussed below.

Molecular Mechanisms of RA Action

The discovery of cellular RA-binding proteins (CRABP) and most importantly the nuclear RA receptors (with the isotypes RARα, β, γ and RXRα, β, γ) has significantly contributed to our understanding of how RA regulates gene expression during embryogenesis [reviewed by Mangelsdorf et al., 1995; Chambon, 1996]. Inactivation of the receptor genes in mice via homologous recombination in ES cells demonstrated that the function of CRABPI and CRABPII is dispensable [Gorry et al., 1994], whereas among the nuclear receptors particularly RXRα/RAR heterodimers seem to be essential for RA signalling [Chambon, 1996; Kastner et al., 1997]. Phenotypic alterations in mice including almost all of the abnormalities observed after postnatal vitamin A deficiency were observed in RAR mutants [Kastner et al., 1995]. The defects regarding heart development in RXRα-deficient mice have been discussed above. The following model for RA signalling is proposed (fig. 6): RA binds to CRABP that interact with the nuclear RA receptors. All-*trans* RA binds to RAR and 9-*cis* RA to RXR and RAR nuclear receptors. Homodimers of RXR or RAR/RXR heterodimers bind via a DNA binding domain to specific sites, the so-called RA response elements. RA response elements are composed of direct repeats of the sequence PuG[G/T]TCA separated by 1, 2 or 5 basepairs, and are located upstream of the target genes. Binding of the nuclear receptors then results in transcription initiation of the target gene.

The pleiotropic effects of RA on different processes during development and differentiation can be explained by regulating gene expression via diverse combinations of RA receptors [Chambon, 1996]. A high level of complexity exists in the RA signalling pathway. Firstly, the different isoforms of RA (e.g. all-*trans* RA and 9-*cis* RA) may be combined in a regulated fashion. Secondly, the two families of RA receptors (RARs and RXRs) are composed of three isotypes (α, β and γ) with several isoforms which can be combined as different RAR/RXR heterodimers. Thirdly, an additional level of complexity is generated by the formation of functionally active heterodimers of RXRs with other nuclear receptors and by interaction of RARs and RXRs with other factors that induce or inhibit gene expression [Chambon, 1996]. Various cell types express different combinations of RA receptors. For example, in EC cells of line P19, the RA receptor RARα, β and γ genes were found to be differentially expressed during differentiation [Kruyt, et al., 1991; Jonk et al., 1992]. During the process of aggregation-mediated endodermal or mesodermal differentiation, no significant alterations

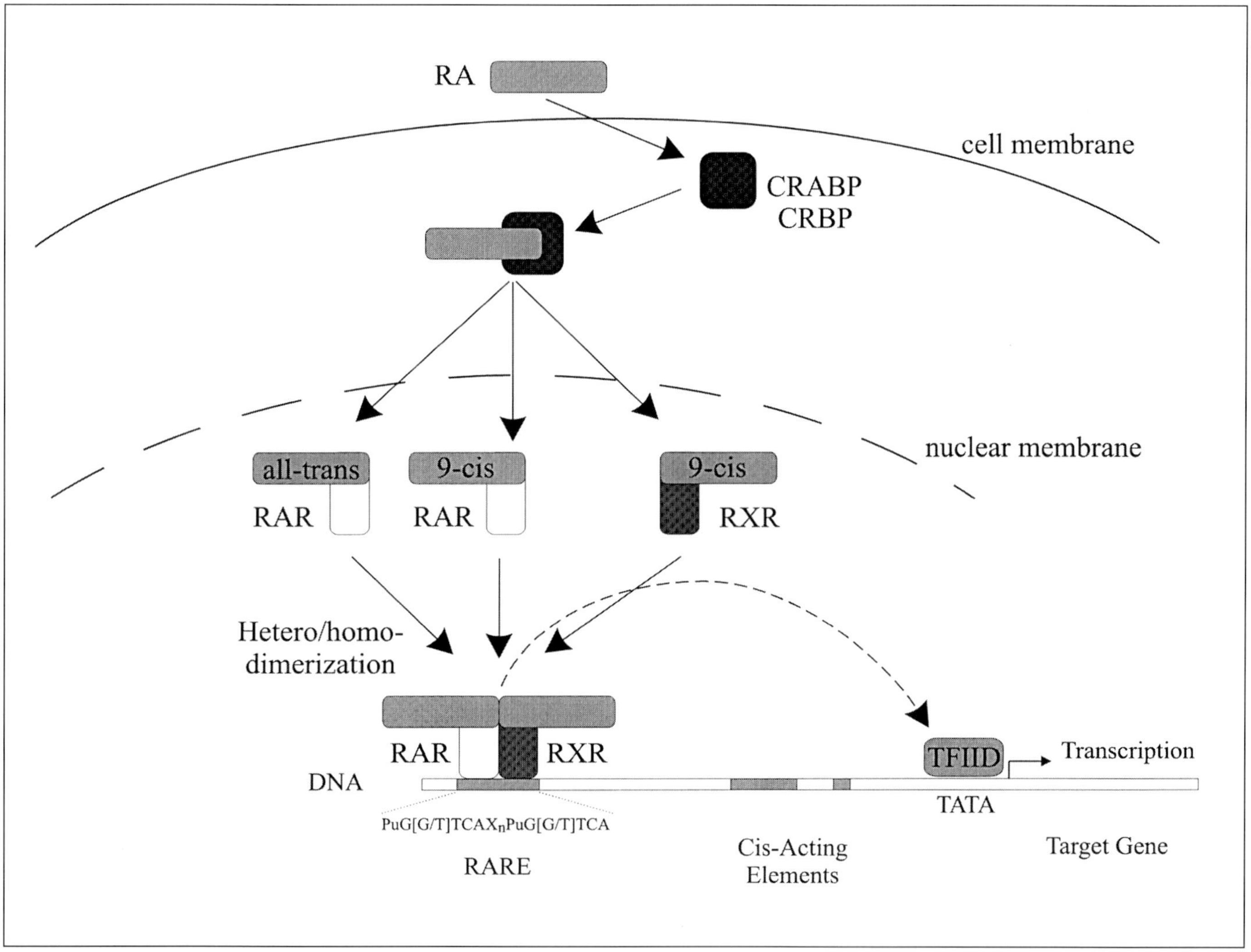

Fig. 6. Mechanism of action of RA and RA-signalling cascades. Via binding to CRABP all-*trans* RA is transferred to RAR, and 9-*cis* RA to RXR and RAR nuclear receptors. RAR/RXR heterodimers (or RXR/RXR homodimers) bind via a DNA binding domain to specific sites, the so-called RA response elements (RARE) upstream of target genes [according to Chambon, 1996].

in mRNA levels for the RA receptors RARα, β or γ could be detected. However, during neuroectodermal differentiation, a strong upregulation of RARα and β, but a rapid downregulation of RARγ mRNA was found, suggesting a role for RARα and RARβ during neuroectodermal differentiation.

As mentioned above, mice deficient for RXRα showed cardiac defects, mainly in the ventricular chamber, whereas mice with single RA receptor gene mutations in either the RARα, β, γ or RXRβ and γ genes did not show any cardiac phenotype [Lee et al., 1997]. However, when different RA receptor gene mutations were combined, specific defects were observed [Lee et al., 1997], suggesting a specific function of different receptor combinations during heart development.

RA might directly act via RA receptor on target genes that play an important role during neuroectodermal differentiation. Using a subtractive hybridization cloning strategy RA-inducible genes were isolated from EC cells of line P19 [Bouillet et al., 1995]. One of these genes, *Stra13* encoding a basic helix-loop-helix protein, is mainly expressed in neuroectoderm during mouse embryogenesis [Boudjelal et al., 1997]. Overexpression of *Stra13* in EC cells of line P19 resulted in induction of neuronal differentiation and altered expression of mesodermal and neuronal marker genes [Boudjelal et al., 1997]. Thus, this gene is a good candidate for functioning as a repressor of mesodermal and an inducer of neuroectodermal differentiation.

With new strategies for gene isolation and identification, RA-induced EC and ES cell-derived differentiated phenotypes may help in the future to identify more RA-specific target genes. Exploring the function of such genes should help to dissect the complex RA signalling pathways and the mechanisms of action of RA on differentiation and embryonic development.

Acknowledgement

This research was supported by the Deutsche Forschungsgemeinschaft (Wo 503/1-3, SFB 366/YE1), and the Fonds der Chemischen Industrie supporting A.M.W.

References

Alles, A.J., K.K. Sulik (1989) Retinoic-acid-induced limb-reduction defects: Perturbation of zones of programmed cell death as a pathogenetic mechanism. Teratology *40:* 163–171.

Bagutti, C., A.M. Wobus, R. Fässler, F.M. Watt (1996) Differentiation of embryonal stem cells into keratinocytes: Comparison of wild-type and β_1 integrin-deficient cells. Dev Biol *179:* 184–196.

Bain, G., D. Kitchens, M. Yao, J.E. Huettner, D.I. Gottlieb (1995) Embryonic stem cells express neuronal properties in vitro. Dev Biol *168:* 342–357.

Blank, R.S., E.A. Swartz, M.M. Thompson, E.N. Olson, G.K. Owens (1995) A retinoic acid-induced clonal cell line derived from multipotential P19 embryonal carcinoma cells express smooth muscle characteristics. Circ Res *76:* 742–749.

Blumberg, B., J. Bolado Jr, T.A. Moreno, C. Kintner, R.M. Evans, N. Papalopulu (1997) An essential role for retinoid signaling in anteroposterior neural patterning. Development *124:* 373–379.

Bollag, W. (1994) Experimental basis of cancer combination chemotherapy with retinoids, cytokines, 1,25-dihydroxyvitamin D_3, and analogs. J Cell Biochem *56:* 427–435.

Boncinelli E, A. Simeone, D. Acampora, F. Mavilio (1991) HOX gene activation by retinoic acid. Trends Genet *7:* 329–334.

Boudjelal, M., R. Taneja, S. Matsubara, P. Bouillet, P. Dolle, P. Chambon (1997) Overexpression of Stra13, a novel retinoic acid-inducible gene of the basic helix-loop-helix family, inhibits mesodermal and promotes neuronal differentiation of P19 cells. Genes Dev *11:* 2052–2065.

Bouillet, P., M. Oulad Abdelghani, S. Vicaire, J.M. Garnier, B. Schuhbaur, P. Dolle, P. Chambon (1995) Efficient cloning of cDNAs of retinoic acid-responsive genes in P19 embryonal carcinoma cells and characterization of a novel mouse gene, Stra1 (mouse LERK-2/Eplg2). Dev Biol *170:* 420–433.

Braun, T., H.-H. Arnold (1994) ES-cells carrying two inactivated myf-5 alleles form skeletal muscle cells: Activation of an alternative myf-5-independent differentiation pathway. Dev Biol *164:* 24–36.

Breier, G., M. Bucan, U. Francke, A.M. Colberg-Poley, P. Gruss (1986) Sequential expression of murine homeo box genes during F9 EC cell differentiation. EMBO J *5:* 2209–2215.

Chambon, P. (1996) A decade of molecular biology of retinoic acid receptors. FASEB J *10:* 940–954.

Chen, Y., L. Huang, A.F. Russo, M. Solursh (1992) Retinoic acid is enriched in Hensen's node and is developmentally regulated in the early chicken embryo. Proc Natl Acad Sci USA *89:* 10056–10059.

Chen, Y., M. Solursh (1992) Comparison of Hensen's node and retinoic acid in seconary axis induction in the early chick embryo. Dev Dyn *195:* 142–151.

Dani, C., A.G. Smith, S. Dessolin, P. Leroy, L. Staccini, P. Villageois, C. Darimont, G. Ailhaud (1997) Differentiation of embryonic stem cells into adipocytes in vitro. J Cell Sci *110:* 1279–1285.

Degos, L., C. Chomienne, M.T. Daniel, R. Berger, H. Dombret, P. Fenaux, S. Castaigne (1990) Treatment of first relapse in acute promyelocytic leukaemia with all-*trans* retinoic acid. Lancet *336:* 1440–1441.

Doetschman, T.C., H. Eistetter, M. Katz, W. Schmidt, R. Kemler (1985) The in vitro development of blastocyst-derived embryonic stem cell lines: Formation of visceral yolk sac, blood islands and myocardium. J Embryol Exp Morphol *87:* 27–45.

Drab, M., H. Haller, R. Bychkov, B. Erdmann, C. Lindschau, H. Haase, I. Morano, F.C. Luft, A.M. Wobus (1997) From totipotent embryonic stem cells to spontaneously contracting smooth muscle cells: A retinoic acid and db-cAMP in vitro differentiation model. FASEB J *11:* 905–915.

Drysdale, T.A., K.D. Patterson, M. Saha, P.A. Krieg (1997) Retinoic acid can block differentiation of the myocardium after heart specification. Dev Biol *188:* 205–215.

Durston, A.J., J.P. Timmermans, W.J. Hage, H.F. Hendriks, N.J. de Vries, M. Heideveld, P.D. Nieuwkoop (1989) Retinoic acid causes an anteroposterior transformation in the developing central nervous system. Nature *340:* 140–144.

Edwards, M.K., J.F. Harris, M.W. McBurney (1983) Induced muscle differentiation in an embryonal carcinoma cell line. Mol Cell Biol *3:* 2280–2286.

Edwards, M.K., M.W. McBurney (1983) The concentration of retinoic acid determines the differentiated cell types formed by a teratocarcinoma cell line. Dev Biol *98:* 187–191.

Eichele, G. (1989) Retinoids and vertebrate limb pattern formation. Trends Genet *5:* 246–251.

Fässler, R., J. Rohwedel, V. Maltsev, W. Bloch, S. Lentini, K. Guan, D. Gullberg, J. Hescheler, K. Addicks, A.M. Wobus (1996) Differentiation and integrity of cardiac muscle cells are impaired in the absence of β_1 integrin. J Cell Sci *109:* 2989–2999.

Fraichard, A., O. Chassande, G. Bilbaut, C. Dehay, P. Savatier, J. Samarut (1995) In vitro differentiation of embryonic stem cells into glial cells and functional neurons. J Cell Sci *108:* 3181–3188.

Gorry, P., T. Lufkin, A. Dierich, C. Rochette-Egly, D. Decimo, P. Dolle, M. Mark, B. Durand, P. Chambon (1994) The cellular retinoic acid binding protein I is dispensable. Proc Natl Acad Sci USA *91:* 9032–9036.

Gruber, P.J., S.W. Kubalak, T. Pexieder, H.M. Sucov, R.M. Evans, K.R. Chien (1996) RXR alpha deficiency confers genetic susceptibility for aortic sac, conotruncal, atrioventricular cushion, and ventricular muscle defects in mice. J Clin Invest *98:* 1332–1343.

Guan, K., J. Rohwedel, A.M. Wobus (1999) Embryonic stem cell differentiation models: Cardiogenesis, myogenesis, neurogenesis, epithelial and vascular smooth muscle cell differentiation in vitro. Cytotechnology *30:* 211–226.

Hemmati-Brivanlou, A., D. Melton (1997) Vertebrate embryonic cells will become nerve cells unless told otherwise. Cell *88:* 13–17.

Hogan, B.L., C. Thaller, G. Eichele (1992) Evidence that Hensen's node is a site of retinoic acid synthesis. Nature *359:* 237–241.

Huang, M.E., Y.C. Ye, S.R. Chen, J.R. Chai, J.X. Lu, L. Zhoa, L.J. Gu, Z.Y. Wang (1988) Use of all-*trams* retinoic acid in the treatment of acute promyelocytic leukemia. Blood *72:* 567–572.

Hunt, P., R. Krumlauf (1991) Deciphering the Hox code: Clues to patterning branchial regions of the head. Cell *66:* 1075–1078.

Jeannotte, L., M. Lemieux, J. Charron, F. Poirier, E.J. Robertson (1993) Specification of axial identity in the mouse: Role of the Hoxa-5 (Hox1.3) gene. Genes Dev *7:* 2085–2096.

Jones-Villeneuve, E.M., M.A. Rudnicki, J.F. Harris, M.W. McBurney (1983) Retinoic acid-induced neural differentiation of embryonal carcinoma cells. Mol Cell Biol *3:* 2271–2279.

Jonk, L.J.C., M.E.J. de Jonge, F.A. Kruyt, C.L. Mummery, P.T. van der Saag, W. Kruijer (1992) Aggregation and cell cycle dependent retinoic acid receptor mRNA expression in P19 embryonal carcinoma cells. Mech Dev *36:* 165–172.

Kastner, P., J.M. Grondona, M. Mark, A. Gansmuller, M. LeMeur, D. Decimo, J.L. Vonesch, P. Dolle, P. Chambon (1994) Genetic analysis of RXR alpha developmental function: Convergence of RXR and RAR signaling pathways in heart and eye morphogenesis. Cell *78:* 987–1003.

Kastner, P., M. Mark, P. Chambon (1995) Nonsteroid nuclear receptors: What are genetic studies telling us about their role in real life? Cell *83:* 859–869.

Kastner, P., M. Mark, N. Ghyselinck, W. Krezel, V. Dupe, J.M. Grondona, P. Chambon (1997) Genetic evidence that the retinoid signal is transduced by heterodimeric RXR/RAR functional units during mouse development. Development *124:* 313–326.

Kessel, M. (1992) Respecification of vertebral identities by retinoic acid. Development *115:* 487–501.

Kessel, M., P. Gruss (1991) Homeotic transformations of murine vertebrae and concomitant alteration of Hox codes induced by retinoic acid. Cell *67:* 89–104.

Kochhar, D.M. (1973) Limb development in mouse embryos. I. Analysis of teratogenic effects of retinoic acid. Teratology *7:* 289–298.

Kolm, P.J., H.L. Sive (1997) Retinoids and posterior neural induction: A reevaluation of Nieuwkoop's two-step hypothesis. Cold Spring Harb Symp Quant Biol *62:* 511–521.

Krumlauf, R. (1994) Hox genes in vertebrate development. Cell *78:* 191–201.

Kruyt, F.A.E., C.E. van den Brink, L.H.K. Defize, M.-J. Donath, P. Kastner, W. Kruijer, P. Chambon, P.T. van der Saag (1991) Transcriptional regulation of retinoic acid receptor beta in retinoic acid-sensitive and -resistant P19 embryocarcinoma cells. Mech Dev *33:* 171–178.

Lammer, E.J., D.T. Chen, R.M. Hoar, N.D. Agnish, P.J. Benke, J.T. Braun, C.J. Curry, P.M. Fernhoff, A.W. Grix Jr., I.T. Lott, J.M. Richard, S.C. Sun (1985) Retinoic acid embryopathy. N Engl J Med *313:* 837–841.

Le Mouellic, H., Y. Lallemand, P. Brulet (1992) Homeosis in the mouse induced by a null mutation in the Hox-3.1 gene. Cell *69:* 251–264.

Lee, R.Y., J. Luo, R.M. Evans, V. Giguere, H.M. Sucov (1997) Compartment-selective sensitivity of cardiovascular morphogenesis to combinations of retinoic acid receptor gene mutations. Circ Res *80:* 757–764.

Lotan, R. (1980) Effects of vitmain A and its analogs (retinoids) on normal and neoplastic cells. Biochim Biophys Acta *605:* 33–91.

Lotan, R. (1996) Retinoids in cancer chemoprevention. FASEB J *10:* 1031–1039.

Maltsev, V.A., J. Rohwedel, J. Hescheler, A.M. Wobus (1993) Embryonic stem cells differentiate in vitro into cardiomyocytes representing sinusnodal, atrial and ventricular cell types. Mech Dev *44:* 41–50.

Maltsev, V.A., A.M. Wobus, J. Rohwedel, M. Bader, J. Hescheler (1994) Cardiomyocytes differentiated in vitro from embryonic stem cells developmentally express cardiac-specific genes and ionic currents. Circ Res *75:* 233–244.

Mangelsdorf, D.J., C. Thummel, M. Beato, P. Herrlich, G. Schutz, K. Umesono, B. Blumberg, P. Kastner, M. Mark, P. Chambon, R.M. Evans (1995) The nuclear receptor superfamily: The second decade. Cell *83:* 835–839.

Marshall, H., S. Nonchev, M.H. Sham, I. Muchamore, A. Lumsden, R. Krumlauf (1992) Retinoic acid alters hindbrain Hox code and induces transformation of rhombomeres 2/3 into a 4/5 identity. Nature *360:* 737–741.

Marshall, H., M. Studer, H. Popperl, S. Aparicio, A. Kuroiwa, S. Brenner, R. Krumlauf (1994) A conserved retinoic acid response element required for early expression of the homeobox gene Hoxb-1. Nature *370:* 567–571.

Morriss-Kay, G.M., N. Sokolova (1996) Embryonic development and pattern formation. FASEB J *10:* 961–968.

Okabe, S., K. Forsberg-Nilsson, A.C. Spiro, M. Segal, R.D. McKay (1996) Development of neuronal precursor cells and functional postmitotic neurons from embryonic stem cells in vitro. Mech Dev *59:* 89–102.

Papalopulu, N., C. Kintner (1996) A posteriorising factor, retinoic acid, reveals that anteroposterior patterning controls the timing of neuronal differentiation in Xenopus neuroectoderm. Development *122:* 3409–3418.

Piccolo, S., Y. Sasai, B. Lu, E.M. De Robertis (1996) Dorsoventral patterning in Xenopus: Inhibition of ventral signals by direct binding of chordin to BMP-4. Cell *86:* 589–598.

Pöpperl, H., M.S. Featherstone (1993) Identification of a retinoic acid response element upstream of the murine Hox-4.2 gene. Mol Cell Biol *13:* 257–265.

Ramirez-Solis, R., H. Zheng, J. Whiting, R. Krumlauf, A. Bradley (1993) Hoxb-4 (Hox-2.6) mutant mice show homeotic transformation of a cervical vertebra and defects in the closure of the sternal rudiments. Cell *73:* 279–294.

Rohwedel, J., K. Guan, W. Zuschratter, S. Jin, G. Ahnert Hilger, D. Fürst, R. Fässler, A.M. Wobus (1998a) Loss of β_1 integrin function results in a retardation of myogenic, but an acceleration of neuronal, differentiation of embryonic stem cells in vitro. Dev Biol *201:* 167–184.

Rohwedel, J., V. Horák, M. Hebrok, E.M. Füchtbauer, A.M. Wobus (1995) M-twist expression inhibits mouse embryonic stem cell-derived myogenic differentiation in vitro. Exp Cell Res *220:* 92–100.

Rohwedel, J., T. Kleppisch, U. Pich, K. Guan, S. Jin, W. Zuschratter, C. Hopf, W. Hoch, J. Hescheler, V. Witzemann, A.M. Wobus (1998b) Formation of postsynaptic-like membranes during differentiation of embryonic stem cells in vitro. Exp Cell Res *239:* 214–225.

Rohwedel, J., V. Maltsev, E. Bober, H.-H. Arnold, J. Hescheler, A.M. Wobus (1994) Muscle cell differentiation of embryonic stem cells reflects myogenesis in vivo: Developmentally regulated expression of myogenic determination genes and functional expression of ionic currents. Dev Biol *164:* 87–101.

Rose, O., J. Rohwedel, S. Reinhardt, M. Bachmann, M. Cramer, M. Rotter, A. Wobus, A. Starzinski-Powitz (1994) Expression of M-cadherin protein in myogenic cells during prenatal mouse development and differentiation of embryonic stem cells in culture. Dev Dyn *201:* 245–259.

Rudnicki, M.A., M.W. McBurney (1987) Cell culture methods and induction of differentiation of embryonal carcinoma cell lines; in Robertson, E.J. (ed): Teratocarcinomas and Embryonic Stem Cells. Oxford, IRL Press, pp 19–49.

Sasai, Y., E.M. De Robertis (1997) Ectodermal patterning in vertebrate embryos. Dev Biol *182:*5–20.

Schmitt, R.M., E. Bruyns, H.R. Snodgrass (1991) Hematopoietic development of embryonic stem cells in vitro: Cytokine and receptor gene expression. Genes Dev *5:* 728–740.

Scott, M.P., J.W. Tamkun, G.W. Hartzell (1989) The structure and function of the homeodomain. Biochim Biophys Acta *989:* 25–48.

Shea, T.B., I. Fischer, V.S. Sapirstein (1985) Effect of retinoic acid on growth and morphological differentiation of mouse NB2a neuroblastoma cells in culture. Brain Res *353:* 307–314.

Simeone, A., D. Acampora, L. Arcioni, P.W. Andrews, E. Boncinelli, F. Mavilio (1990) Sequential activation of *HOX2* homeobox genes by retinoic acid in human embryonal carcinoma cells. Nature *346:* 763–766.

Simon, M.C., L. Pevny, M.V. Wiles, G. Keller, F. Costantini, S.H. Orkin (1992) Rescue of erythroid development in gene targeted GATA-1-mouse embryonic stem cells. Nature Genet *1:* 92–98.

Sive, H.L., B.W. Draper, R.M. Harland, H. Weintraub (1990) Identification of a retinoic acid-sensitive period during primary axis formation in *Xenopus laevis.* Genes Dev *4:* 932–942.

Soudais, C., M. Bielinska, M. Heikinheimo, C.A. MacArthur, N. Narita, J.E. Saffitz, M.C. Simon, J.M. Leiden, D.B. Wilson (1955) Targeted mutagenesis of the transcription factor GATA-4 gene in mouse embryonic stem cells disrupts visceral endoderm differentiation in vitro. Development *121:* 3877–3888.

Strickland, S., Mahdavi (1978) The induction of differentiation in teratocarcinoma stem cells by retinoic acid. Cell *15:* 393–403.

Strübing, C., G. Ahnert Hilger, J. Shan, B. Wiedenmann, J. Hescheler, A.M. Wobus (1995) Differentiation of pluripotent embryonic stem cells into the neuronal lineage in vitro gives rise to mature inhibitory and excitatory neurons. Mech Dev *53:* 275–287.

Sucov, H.M., E. Dyson, C.L. Gumeringer, J. Price, K.R. Chien, R.M. Evans (1994) RXR alpha mutant mice establish a genetic basis for vitamin A signaling in heart morphogenesis. Genes Dev *8:* 1007–1018.

Tabin, C.J. (1991) Retinoids, homeoboxes, and growth factors: Toward molecular models for limb development. Cell *66:* 199–217.

Taylor, S.M., P.A. Jones (1979) Multiple new phenotypes induced in 10T1/2 and 3T3 cells treated with 5-azacytidine. Cell *17:* 771–779.

Thomas, K.R., M.R. Capecchi (1987) Site-directed mutagenesis by gene targeting in mouse embryo-derived stem cells. Cell *51:* 503–512.

Wiles, M.V. (1993) Embryonic stem cell differentiation in vitro. Methods Enzymol *225:* 900–918.

Wilson, J.G., C.B. Roth, J. Warkany (1953) An analysis of the syndrome of malformations induced by maternal vitamin A deficiency. Effects of restoration of vitamin A at various times during gestation. Am J Anat *92:* 189–217.

Wilson, J.G., J. Warkany (1949) Aortic arch and cardiac anomalies in the offspring of vitamin A deficient rats. Am J Anat *85:* 113–155.

Wobus, A.M., K. Guan (1998) Embryonic stem cell-derived cardiac differentiation: Modulation of differentiation and 'loss-of-function' analysis in vitro. Trends Cardiovasc Med *8:* 64–74.

Wobus, A.M., K. Guan, J. Shan, M.C. Wellner, J. Rohwedel, J. Guanju, B. Fleischmann, H.A. Katus, J. Hescheler, W.M. Franz (1997) Retinoic acid accelerates embryonic stem cell-derived cardiac differentiation and enhances development of ventricular cardiomyocytes. J Mol Cell Cardiol *29:* 1525–1539.

Wobus, A.M., T. Kleppisch, V. Maltsev, J. Hescheler (1994a) Cardiomyocyte-like cells differentiated in vitro from embryonic carcinoma cells P19 are characterized by functional expression of adrenoceptors and Ca^{2+} channels. In Vitro Cell Dev Biol Anim *30A:* 425–434.

Wobus, A.M., J. Rohwedel, V. Maltsev, J. Hescheler (1994b) In vitro differentiation of embryonic stem cells into cardiomyocytes or skeletal muscle cells is specifically modulated by retinoic acid. Rouxs Arch Dev Biol *204:* 36–45.

Wobus, A.M., G. Wallukat, J. Hescheler (1991) Pluripotent mouse embryonic stem cells are able to differentiate into cardiomyocytes expressing chronotropic responses to adrenergic and cholinergic agents and Ca^{2+} channel blockers. Differentiation *48:* 173–182.

Wolf, G. (1996) A history of vitamin A and retinoids. FASEB J *10:* 1102–1107.

Yu, V.C., C. Delsert, B. Andersen, J.M. Holloway, O.V. Devary, A.M. Naar, S.Y. Kim, J.M. Boutin, C.K. Glass, M.G. Rosenfeld (1991) RXR beta: A coregulator that enhances binding of retinoic acid, thyroid hormone, and vitamin D receptors to their cognate response elements. Cell *67:* 1251–1266.

Cells Tissues Organs

Cells Tissues Organs 1999;165:203–211

Accepted after revision: July 8, 1999

Embryotoxicity Screening Using Embryonic Stem Cells in vitro: Correlation to in vivo Teratogenicity

Gabriele Scholz[a] Ingeborg Pohl[a] Elke Genschow[b] Martina Klemm[a] Horst Spielmann[a]

[a]Center for Documentation and Evaluation of Alternative Methods to Animal Experiments (ZEBET) at the [b]Federal Institute for Health Protection of Consumers and Veterinary Medicine (BgVV), Berlin, Germany

Key Words

Differentiation in vitro · Embryonic stem cell test · Cardiomyocytes · Cytotoxicity · Embryotoxicity · Prediction model

Abstract

Blastocyst-derived pluripotent embryonic stem (ES) cells of the mouse can be induced to differentiate in culture into a variety of cell types, including cardiac muscle cells. The embryonic stem cell test that makes use of the differentiation of ES cells into cardiomyocytes in a standardized in vitro model was developed to offer an alternative method to comprehensive in vivo studies in reproductive toxicology about toxic effects of chemicals. ES cells of the mouse cell line D3 are investigated for their preserved capability to differentiate following drug exposure, and both ES cells and differentiated fibroblast cells of the mouse cell line 3T3 are comparatively analyzed for effects on viability. The following endpoints are used to classify the embryotoxic potential of chemicals into three classes of in vitro embryotoxicity (non-, weakly or strongly embryotoxic). These endpoints are: (1) the inhibition of differentiation of ES cells into cardiomyocytes after 10 days of treatment, and the decrease of viability (cytotoxicity) of (2) 3T3 cells and (3) ES cells after 10 days of treatment, determined by a 3-(4,5-dimethylthiazol-2yl)-2,5-diphenyl tetrazolium bromide (MTT) test. 50% inhibition concentrations for differentiation (ID_{50}) and cytotoxicity ($IC_{50}D3$ and $IC_{50}3T3$) are calculated from concentration-response curves. Applying linear analysis of discriminance, a biostatistical prediction model (PM) was developed. This procedure identified three variables, the $lg(IC_{50}D3)$, the $lg(IC_{50}3T3)$ and the relative distance between $IC_{50}3T3$ and ID_{50}, that improved the separation of the three classes of embryotoxicity compared to the pre-

Abbreviations used in this paper

Asc	ascorbic acid
DPH	diphenylhydantoin
EB	embryoid body
ES cells	embryonic stem cells
EST	embryonic stem cell test
5-FU	5-fluorouracil
LIF	leukemia inhibitory factor
MM	micromass
MTT	3-(4,5-dimethylthiazol-2yl)-2,5-diphenyl tetrazolium bromide
PM	prediction model
RA	retinoic acid
WEC	whole embryo culture

KARGER

Fax + 41 61 306 12 34
E-Mail karger@karger.ch
www.karger.com

1422–6405/99/1654–0203$17.50/0

Accessible online at:
www.karger.com/journals/cto

Dr. Gabriele Scholz
ZEBET at the BgVV, Diedersdorfer Weg 1
D–12277 Berlin (Germany)
Tel. +49 30 8412 2270, Fax +49 30 8412 2958
E-Mail zebet@bgvv.de

diction model that was originally proposed after test development. Unlike the orginal PM, the improved PM incorporates as one variable the relative distance between $IC_{50}3T3$ and ID_{50}, instead of the ratio $ID_{50}/IC_{50}D3$ that was used previously.

Introduction

Developing organisms undergo rapid and complex changes within a relatively short period of time. Therefore, in mammals any agent administered during pregnancy to the mother under appropriate conditions of time and dosage may interfere with embryonic development and induce embryo lethality, growth retardation or teratogenic effects, with structural and functional abnormalities in the offspring. Persistent lesions, e.g. general growth retardation or delayed organ growth, are defined as embryotoxic effects. Currently, according to the great number of existing industrial chemicals that are commercially available, screening tests or multigeneration studies must be conducted to provide information on the toxic effects on specific elements of the highly complicated reproductive cycle according to guidelines of the OECD [OECD Test Guidelines No. 414, 1981; No. 415, 1983; No. 416, 1983; No. 421, 1995; No. 422, 1996]. For chemicals used as drugs, segment studies have to be conducted covering preconceptional exposure as well as pre- and postnatal development including the lactation period [guidelines of the International Conference on Harmonisation, ICH, 1993].

These in vivo protocols are time-consuming, expensive and they are carried out on large numbers of laboratory animals. One important feature of the in vivo test protocols is the assessment of maternal toxicity in comparison to adverse effects in the offspring. Today, these two important aspects can to some extent be covered by in vitro approaches. In developmental toxicology, many in vitro alternatives to testing in animals have been developed using a wide spectrum of cell and tissue cultures, e.g. permanent cell lines, cultures of primary embryonic cells and cultures of nonmammalian tissue and mammalian embryos [for a review, see Brown et al., 1995; Spielmann, 1998]. For example, in the rat limb bud micromass (MM) assay, effects on the viability of primary limb bud cell cultures (cytotoxicity) are compared to effects on the differentiation of these cells into chondrocytes [Flint and Orton, 1984]. In the postimplantation rat whole embryo culture (WEC) assay, both general growth retardation and specific malformations of the cultivated embryos are assessed [for a review, see Spielmann, 1998].

One of the more recently developed in vitro approaches is based on blastocyst-derived pluripotent embryonic stem (ES) cells of the mouse. In contrast to the WEC and MM tests, the embryonic stem cell test (EST) [Spielmann et al., 1997] has the advantage of using established cell lines without the need to sacrifice pregnant animals.

Since cultivation and maintenance of ES cells in the undifferentiated state was first reported by Evans and Kaufman [1981], cultivation and differentiation of ES cells is today a widely applied method in mammalian developmental biology. ES cells can be genetically manipulated to generate transgenic or 'knockout' mice [Thomas and Capecchi, 1987], and in vitro cell culture models were established to study myogenesis, angiogenesis, hematopoiesis, neurogenesis and cardiogenesis in the mouse [Wiles and Keller, 1991; Wobus et al., 1991; Heuer et al., 1994; Maltsev et al., 1994; Rohwedel et al., 1994; Strübing et al., 1995; Kolossov et al., 1998].

The differentiation of ES cells into cardiac cells [Doetschman et al., 1985] has been used in investigations on prenatal pharmacology, electrophysiology and molecular genetics [Wobus et al., 1991; Maltsev et al., 1993; Metzger et al., 1996; Kolossov et al., 1998]. The EST developed by Spielmann and coworkers [1997] was designed to adopt the ES cell cultures in embryotoxicity studies by using an endpoint related to prenatal differentiation in vivo, the differentiation of ES cells into contracting cardiac muscle cells. The inhibition of this specific differentiation by embryotoxic agents was compared to cytotoxic effects in the ES cells [Laschinski et al., 1991]. To improve the predictive potential of this in vitro embryotoxicity test, in the EST, the influence of potentially embryotoxic chemicals on the viability of ES cells and on their capability to differentiate in vitro [Wobus et al., 1991; Rohwedel et al., 1994; Spielmann et al., 1997] is examined in the permanent mouse ES cell line D3 and compared to the effect on viability of differentiated fibroblast cells of the line 3T3.

Despite the use of some established alternative embryotoxicity tests like the MM test and the WEC test in the drug and chemical industry for screening new chemicals that are structurally related to established embryotoxic chemicals, none of these in vitro tests has to date been sufficiently validated for regulatory purposes according to the recommendations of the European Centre for the Validation of Alternative Methods (ECVAM) [Balls et al., 1995]. Therefore, in developmental toxicity testing there is a strong demand for validated in vitro tests using mammalian embryos as well as primary cultures of embryonic cells and permanent cell lines. The limiting factors for in vitro screening are obvious when taking into account the complexity of normal differentiation and development of mammalian embryos. Nevertheless, three of the most promising in vitro tests are currently undergoing formal validation in a European ring trial: the WEC test, the MM test and the EST. This validation study is aimed at predicting the embryotoxic potential of a set of test chemicals characterized by high-quality in vivo embryotoxicity data in laboratory animals and humans [Scholz et al., 1998; Genschow et al., 1999].

In this study the performance in the EST to predict the embryotoxic potential of test chemicals is described and we report on the development of an improved prediction model for the EST, derived from results obtained in our laboratory in a prevalidation trial that preceded the currently running validation study. The importance of sound biometrical prediction models as a means for an objective evaluation and classification of the performance of in vitro tests in validation studies is discussed.

Test Procedure of the EST

BALB/c 3T3 cells (clone 31, ICN Flow, Eschwege, Germany) are routinely maintained in DMEM supplemented with 10% FCS, 4 m*M* glutamine, 50 U/ml penicillin and 50 mg/ml streptomycin. ES cells (line D3, a donation from Prof. R. Kemler, MPI, Freiburg, Germany) are maintained in DMEM supplemented with 20% FCS, 2 m*M* glutamine, 50 U/ml penicillin, 50 mg/ml streptomycin, 1% nonessential amino acids, 0.1% β-mercaptoethanol and 1,000 U/ml leukemia inhibitory factor (LIF) [Williams et al., 1988].

To determine cytotoxic effects on ES and 3T3 cells, 3-(4,5-dimethylthiazol-2yl)-2,5-diphenyl tetrazolium bromide (MTT) tests are performed in the absence of LIF as described previously [Spielmann et al., 1997]. Briefly, 500 cells are seeded into each well of a 96-well microtiter plate and grown in the presence of a concentration range of the test chemical. A negative control containing solvent diluted in medium is also included. After 10 days of culture with two changes of medium (containing the appropriate concentration of test chemical) on days 3 and 5, the viability of the cells is determined using an MTT test, which is based on the capacity of mitochondrial dehydrogenase enyzmes in living cells to convert the yellow substrate MTT into a dark blue formazan product, which is detected quantitatively using a microplate ELISA reader. Cytotoxicity is expressed as the concentration decreasing viability to 50% of the control level (IC_{50}3T3 and IC_{50}D3; determined from a concentration-response curve).

To detect effects on the differentiation of ES cells into cardiomyocytes, a differentiation assay is performed as described previously [Spielmann et al., 1997]. Briefly, 750 cells in drops of 20 μl are seeded (in the absence of LIF to allow differentiation) into the lid of a culture dish and grown for 3 days in the 'hanging drop' culture in the presence of a concentration range of the test chemical. During this period the cells will form aggregates called embryoid bodies (EBs). After 3 days of 'hanging drop' culture the EBs are transferred to bacterial petri dishes containing the appropriate concentration of the test chemical for another 2 days. Bacterial petri dishes are used to avoid adherence and outgrowth of the EBs during this stage of the culture. On day 5 EBs are seeded separately into wells of a 24-well tissue culture plate (containing the appropriate concentration of test chemical) to allow adherence and outgrowth of the EBs and development of contracting cardiomyocytes. Differentiation is assessed by microscopic inspection of the EBs on day 10. The percentage of wells of each plate in which contracting cardiomyocytes have developed is determined and compared to the percentage of wells in which cardiomyocytes developed on the solvent control plate. The inhibition of differentiation (ID_{50}) is expressed as the concentration of the test chemical inhibiting the development into contracting cardiomyocytes by 50% (calculated from the concentration-response curve).

Classification of Embryotoxic Chemicals

In a recent prevalidation study that was conducted according to the recommendations of the European Centre for the Validation of Alternative Methods (ECVAM) [Curren et al., 1995; Southee and Curren, 1997], chemicals assigned to the three classes of embryotoxicity were tested under blind conditions in order to prove that the assay is reproducible and transferable to other laboratories and that it is able to predict the embryotoxic potential of a given set of chemicals.

In embryotoxicity testing the selection of test chemicals is particularly difficult, since the number of chemicals which are backed by high-quality in vivo data is rather limited. For the prevalidation trial, only test chemicals were used for which in vivo data of a sufficiently high quality were available from both testing in animals and human pregnancy. These were selected from a list of

Table 1. Set of test chemicals

Embryotoxicity in vivo	Test chemical	CAS No.	In vivo activity species	Use
Class 1: nonembryotoxic	Penicillin G	69-57-8	–	Antibiotic
	Isoniacide	54-85-3	–	Tuberculostatic
	Ascorbic acid	134-03-2	–	Vitamin
Class 2: weakly embryotoxic	Diphenylhydantoin	630-93-3	Human, animal	Anticonvulsant
	Caffeine	58-08-2	Human, animal	Central stimulant, drug
	Dexamethasone	50-02-2	Human, animal	antiasthmatic, glucocorticoid
Class 3: strongly embryotoxic	Cytosine arabinoside	69-74-9	Human, animal	Cytostatic
	All-*trans*-RA	302-79-4	Human, animal	Acne therapy
	Hydroxyurea	127-07-1	Animal	Cytostatic
	5-Fluorouracil	51-21-8	Animal	Cytostatic

Characteristics of the test chemicals used in the study are described: 9 test chemicals and the positive control 5-FU. The embryotoxicity in vivo and their use in humans are given. Test chemicals with known embryotoxic potential in vivo were selected from a published list recommended by the US Teratology Society for the validation of in vitro embryotoxicity tests [Smith et al., 1983]. All chemicals were purchased from Sigma-Aldrich Chemie (Steinheim, Germany). Chemicals were dissolved according to a strategy developed for dissolving of coded chemicals, which has been included in the standard protocol. Briefly, a hierarchical approach was used to find an appropriate concentration in a suitable solvent, starting with PBS or medium as solvent, then 50% (volv/vol) ethanol in PBS, then DMSO, and finally pure ethanol.

chemicals recommended by the US Teratology Society for in vitro teratogenesis test validation [Smith et al., 1983]. Nine test chemicals and a positive control of known embryotoxic potential in vivo were selected for testing during the prevalidation study (table 1), mainly consisting of drugs which have been used in human pregnancy, e.g. in case of bacterial infections or the therapy of cancer. The prevalidation trial was conducted in two European and one US laboratory, yet the exact results obtained during this interlaboratory study are reported elsewhere [Scholz et al., 1999]. This report focuses exclusively on the results of our laboratory, since they were used as the basis for the development of an improved PM.

Most of the chemicals tested (table 1) can be categorized in vitro roughly according to typical concentration-response curves obtained for the differentiation of ES cells and for the viability of ES and 3T3 cells in the presence of a concentration range of the test chemical.

Strongly embryotoxic chemicals usually inhibit differentiation of ES cells at very low concentrations and many of them also show a high cytotoxic potential at low concentrations both to 3T3 mouse fibroblast cells and ES cells, e.g. cytostatic drugs [5-fluorouracil (5-FU); see also fig. 1]. Chemicals characterized by a high teratogenic potential, e.g. retinoic acid (RA), are inhibiting differentiation at very low concentrations, while cytotoxic effects are apparent at concentrations that are several orders of magnitude higher (fig. 2a). On the other hand, weakly embryotoxic chemicals are effective in the EST at intermediate concentrations, where the differentiation of ES cells into cardiomyocytes is more sensitive than growth inhibition, e.g. in case of the anticonvulsant drug diphenylhydantoin (DPH) (fig. 2b). Chemicals that do not exhibit any embryotoxic potential, e.g. the sweetener saccharine or the antibiotic penicillin G, are neither inhibiting growth nor differentiation of ES cells at concentrations of more than 500 μg/ml. Other nonembryotoxic chemicals, however, as for instance the vitamin and food additive ascorbic acid (ASC), exert cytotoxic effects at relatively low concentrations on differentiated 3T3 cells, whereas cell differentiation and/or viability of ES cells are less sensitive (fig. 2c).

Improved Prediction Model

A biostatistical evaluation was conducted on the limited set of 9 chemicals and the positive control 5-FU that were tested in our laboratory during prevalidation of the EST [Scholz et al., 1999]. In the process of validation, in vitro tests have to prove their reliability and relevance. For the first requirement, the reliability, individual labo-

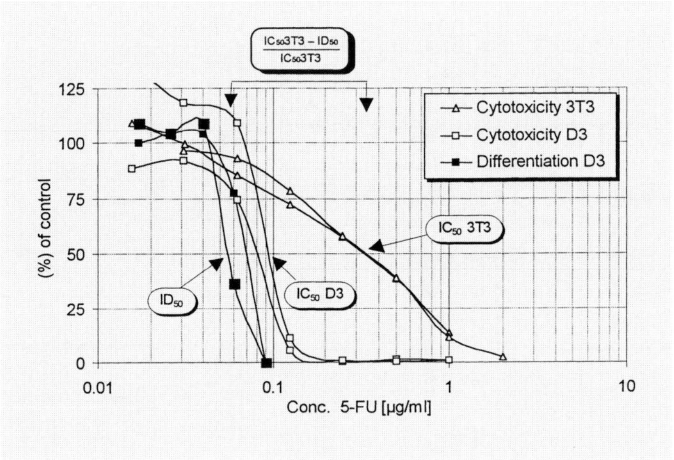

1

Fig. 1. Endpoints and variables used in the improved PM. Diagram shows concentration-response curves for the three endpoints: cytotoxicity of D3 and 3T3 cells and inhibition of differentiation of D3 cells. The curves were obtained from two individual experiments using the positive control chemical 5-fluorouracil (5-FU). Endpoints ($IC_{50}D3$, $IC_{50}3T3$, ID_{50}) and the variable:

$$\left(\frac{IC_{50}3T3 - ID_{50}}{IC_{50}3T3}\right)$$

are marked by arrows.

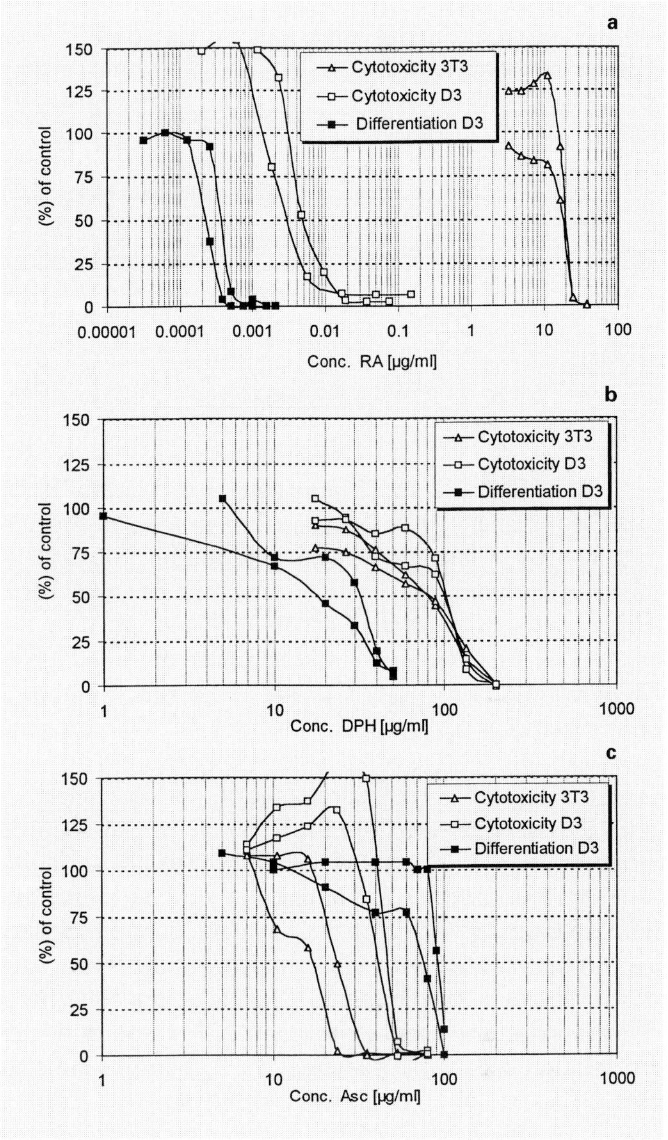

2

Fig. 2. Typical concentration-response curves. **a** Concentration-response curves for the three endpoints of the EST: cytotoxicity of D3 and 3T3 cells and inhibition of differentiation of D3 cells. The curves were obtained from two individual experiments using the strongly embryotoxic chemical retinoic acid (RA). **b** Concentration-response curves of two individual experiments using the weakly embryotoxic diphenylhydantoin (DPH). **c** Concentration-response curves of two individual experiments using the nonembryotoxic ascorbic acid (Asc).

ratories verify a sufficiently high level of exactness by repeating results independently [Balls et al., 1990, 1995]. The second requirement, the relevance of an in vitro test for a specific application, is estimated by applying a biostatistically based prediction model (PM). For developing a scientifically acceptable in vitro embryotoxicity test, the PM should allow to discriminate between three classes of embryotoxicity; non-, weakly and strongly embryotoxic according to in vivo data.

The PM originally developed for the EST [Spielmann et al., 1997] is based on a mathematical calculation, the linear discriminant analysis [Norušis, 1994], taking into account three endpoints used to discriminate between the three embryotoxicity classes: cytotoxicity of 3T3 fibroblasts and ES cells and differentiation of ES cells into contracting cardiomyocytes after 10 days of culture.

The concentration-response curves were analyzed in order to determine appropriate endpoints that allow to discriminate between the three classes of embryotoxic chemicals. First, 50% inhibition concentrations (IC_{50}) were calculated from the concentration-response curves for cytotoxicity of D3 cells and 3T3 fibroblasts and for differentiation of D3 cells, in order to evaluate whether individual endpoints alone sufficiently allow to discriminate between the three classes of embryotoxicity.

Table 2. Improved PM for the EST: linear discriminant functions I, II and III and classification criteria (class 1–3)

Function I	$5.92\ \lg(IC_{50}3T3) + 3.50\ \lg(IC_{50}D3) - 5.31\ \frac{IC_{50}3T3 - ID_{50}}{IC_{50}3T3} - 15.7$
Function II	$3.65\ \lg(IC_{50}3T3) + 2.39\ \lg(IC_{50}D3) - 2.03\ \frac{IC_{50}3T3 - ID_{50}}{IC_{50}3T3} - 6.85$
Function III	$-0.125\ \lg(IC_{50}3T3) - 1.92\ \lg(IC_{50}D3) + 1.50\ \frac{IC_{50}3T3 - ID_{50}}{IC_{50}3T3} - 2.67$
Class 1: nonembryotoxic	If I > II and I > III
Class 2: weakly embryotoxic	If II > I and II > III
Class 3: strongly embryotoxic	If III > I and III > II

Second, figures 1 and 2 indicate that the variability of the IC_{50} is higher compared to the 25% inhibition concentrations (IC_{25}). Therefore, the discriminating potential of the two endpoints (IC_{25}, IC_{50}) had to be evaluated. Although the IC_{25} showed a better performance in the analysis of discriminance – and therefore provided a better discrimination – the IC_{50} was chosen as the appropriate endpoint for the following reasons. When testing chemicals under blind conditions, it is sometimes difficult to completely cover the entire range of the concentration-response curve between 100 and 0% both for viability and differentiation. Furthermore, given the intrinsic variability of the assay and the limited number of concentrations that can be tested within one experiment, the dilution factor should be sufficiently small to achieve the most exact measurement of the endpoint. Therefore, the concentration-response curves may not reach 25% viability or differentiation in each case, whereas the 50% inhibition concentrations are more easily met. Thus, the 50% viability and differentiation (IC_{50}, ID_{50}) was applied as endpoint in the linear discriminant analysis in the improved prediction model.

Yet the endpoint value (IC_{50}, ID_{50}) cannot be used directly as a variable in the linear analysis of discriminance. In order to convert the numerical value of the endpoint into a linear model the logarithm of the IC_{50} is calculated. The term 'endpoint' is used only for the 50% cytotoxic concentrations ($IC_{50}3T3$, $IC_{50}D3$) and the 50% differentiation concentration (ID_{50}), that are directly calculated from the concentration-response curve. The term 'variable' is applied to all mathematical calculations (logarithm, relative distance) of the original endpoints. The improvement of the PM described here is based on the same endpoints as the original PM, but uses a slightly different variable.

In order to include occasionally divergent test results on viability and differentiation into the model, the relative distance between two of the endpoints ($IC_{50}3T3$, $IC_{50}D3$) was used as a variable (fig. 1). This variable is independent from the absolute concentration values, since this information is already given by the two other variables, the $\lg(IC_{50}3T3)$ and the $\lg(IC_{50}D3)$ values.

The in vitro test provided a total of 9 different experimental variables [$\lg(IC_{50}3T3)$; $\lg(IC_{50}D3)$; $\lg(ID_{50})$; $\lg(IC_{25}3T3)$; $\lg(IC_{25}D3)$; $\lg(ID_{25})$; relative distance between $IC_{50}3T3$ and $IC_{50}D3$; relative distance between $IC_{50}3T3$ and ID_{50}, and relative distance between $IC_{50}D3$ and ID_{50}], each of which may contribute to distinguishing between the three groups of embryotoxic chemicals. Consequently, a stepwise selection of variables was performed [Bortz et al., 1993; Norušis, 1994; Backhaus et al., 1996]. This procedure identifies the best variable to discriminate between the selected classes. Subsequently, in a stepwise fashion, each of remaining variables is separately added to the model and rejected again, if it does not significantly improve the separation of three classes of embryotoxicity. Three variables were accepted in the analysis of discriminance which improved the distinction between the three embryotoxic classes compared to the original model. The linear discriminant functions of the PM for the EST incorporating the variables selected are shown in table 2.

The following procedure is ensued to classify the chemicals according to the improved PM: 50% inhibition concentrations ($IC_{50}3T3$, $IC_{50}D3$, ID_{50}) and the relative distance between $IC_{50}3T3$ and ID_{50} are determined and employed in the three linear discriminant functions. A chemical is classified as not embryotoxic, if the result of equation I exceeds the results of equations II and III. A chemical is classified as weakly embryotoxic, if the result of equation II exceeds the results of equations I and III. Finally, if the result of equation III exceeds the results of

Table 3. Classification of the test chemicals

Embryotoxicity in vivo	Number of chemicals	Number of cases	Prediction in vitro		
			1	2	3
Class 1: nonembryotoxic	3	9	*9 (100.0)*	0 (0)	0 (0)
Class 2: weakly embryotoxic	3	9	1 (11.1)	*8 (88.9)*	0 (0)
Class 3: strongly embryotoxic	4	12	0 (0)	1 (8.3)	*11 (91.7)*

The 10 chemicals were repeatedly tested, resulting in 30 individual experiments (number of cases). Correct classifications are italicized. 100% of the nonembryotoxic chemicals, 89% of the weakly embryotoxic chemicals and 92% of the strongly embryotoxic chemicals were classified correctly. In total 94% of all cases were classified correctly. To identify the endpoints, linear discriminant analysis was used [Bortz, 1993; SPSS, stepwise discriminant analysis, Norušis, 1994; Backhaus et al., 1996]. Figures in parentheses represent percentage.

equations I and II, the chemical is classified as strongly embryotoxic.

The classification of the so-called 'learning sample' according to the improved PM provided 94% correct classifications (table 3). However, a model usually fits the learning sample better than it will fit new chemicals. Thus the percentage of correctly classified cases is an overly optimistic estimation and a lower rate for correct classifications must be expected when new chemicals are evaluated with the model. Currently, a set of 20 chemicals is tested in a validation study of three in vitro embryotoxicity tests, including the EST. The results of this study will be used to evaluate the improved PM.

EST Detects Various Types of Teratogenic/Embryotoxic Chemicals

Teratogenic chemicals are distributed in numerous types of chemicals and function via diverse pathways, either highly specific receptor signalling pathways, or more general mechanisms. Although the molecular functions of several teratogens are the subject of intensive research, such as the RA receptor pathway or the arylhydrocarbon receptor pathway, the mechanisms of how adverse effects are affecting normal mammalian development are highly complex and have not sufficiently been identified. Therefore, establishing a highly specific, mechanistic test is impossible to date. Taking into account known mechanisms of embryotoxicity we have used the differentiation of ES cells into contracting cardiomyocytes as a means to measure differentiation and early mammalian development. Our results show that a variety of strong and weak teratogens can correctly be identified in the EST, even chemicals which do not specifically act on early heart development in vivo. For instance, in vivo exposure of early postimplantation rodent embryos to all-*trans*-RA frequently results in craniofacial defects and defects of the central nervous system. Exposure at later developmental stages is often associated with limb and genitourinary defects [Kochhar, 1997, for effects of RA on cardiac development, see Rohwedel et al., 1999]. Nevertheless, RA used under our treatment conditions is a highly potent inhibitor of cardiomyocyte differentiation in vitro, and we did not obtain false-negative results in the EST so far during test development and prevalidation. However, it must be kept in mind that teratogens which specifically interfere with the development of other tissues, e.g. the palate or nervous system, may not be detected in the EST.

The EST and the improved PM for the EST are based on the different responsiveness of pluripotent ES cells and differentiated fibroblast cells to teratogenic/embryotoxic agents. This difference, especially with respect to the cytotoxicity tests with the two cell lines, may to some extent be influenced by the different media requirements and cell growth characteristics of D3 and 3T3 cells. On the one hand, media for 3T3 cells contain less serum (10% FCS) than media for the D3 cells (20% FCS). Since it is known that chemicals can have different affinities for serum proteins, the concentration of free and active compounds may be different in the presence of different serum concentrations, and therefore a higher sensitivity of 3T3 cells in the cytotoxicity test would be expected. On the other hand, 3T3 cells have a longer doubling time than D3 cells and display contact inhibition, which could also explain a higher sensitivity of D3 cells and cytotoxicity tests in some cases.

Since the EST is the only mammalian in vitro alternative to animal testing in developmental toxicology to date which is based on established cell lines and does not require pregnant animals, this test offers the possibility to be optimized for automated high throughput screening systems in toxicity testing of drugs and other chemicals for regulatory purposes. An earlier and slightly different approach to using ES cells in an in vitro embryotoxicity test, which used colony formation as a morphological endpoint of differentiation, showed a very poor prediction and had not been developed further [Newell and Beedles, 1994].

Further Strategies in the Improvement of the EST

Attempts to further optimize the EST test protocol are currently under way in order to identify endpoints of differentiation other than microscopic evaluation of contracting areas. This aim may be achieved by using promoter/reporter gene expression techniques and stably transfected ES cells which are expressing, for example, green fluorescent protein under the control of a developmentally regulated promoter, such as the cardiac α-actin promoter which is activated during cardiac muscle cell development [Kolossov et al., 1998]. Another approach may focus on endogenous gene expression by RT-PCR methods, for which protocols and primers exist to detect early markers of gene expression in early development as well as gene expression patterns in developing EB with a very limited amount of ES cells [Wiles, 1993]. Such molecular techniques would offer the advantage of replacing the subjective evaluation through an experienced person by an objective measurement that can be automated, which is an important prerequisite for high throughput screening systems. Furthermore, the assay duration could be reduced, from 10 to e.g. 7 or even 5 days.

The inclusion of other endpoints in addition to cardiac muscle cell development is another important consideration in order to reduce the risk of obtaining false-negative results. For instance, early hematopoiesis and blood cell development can be induced in ES cell cultures under defined culture conditions and by the addition of specific growth factors.

Thus, there are several promising strategies to adapt or extend in vitro ES cell differentiation techniques in order to reduce animal testing in reproductive toxicology and at the same time to gain further insight into mechanisms of action of teratogens on the cellular and molecular level.

Acknowledgements

This study was funded by the European Centre for the Validation of Alternative Methods (ECVAM) at the EU Joint Research Centre (JRC) in Ispra, Italy, through a subcontract of JRC Grant No. 11279-95010FIED ISP GB awarded to Microbiological Associates Ltd., Stirling (UK).

References

Anon (1981) OECD Guideline for Testing of Chemicals, No 414: Teratogenicity. Paris, Organisation for Economic Co-Operation and Development.

Anon (1983) OECD Guideline for Testing of Chemicals, No 415: One Generation Reproduction Toxicity Study. Paris, Organisation for Economic Co-Operation and Development.

Anon (1983) OECD Guideline for Testing of Chemicals, No 416: Two Generation Reproduction Toxicity Study. Paris, Organisation for Economic Co-Operation and Development.

Anon (1995) OECD Guideline for Testing of Chemicals, Draft Guideline No 421: Reproduction/Developmental Toxicity Screening Test. Paris, Organisation for Economic Co-Operation and Development.

Anon (1996) OECD Guideline for Testing of Chemicals, Draft Guideline No 422: Combined Repeated Dose Toxicity Study with the Reproduction/Developmental Toxicity Screening Test. Paris, Organisation for Economic Co-Operation and Development.

Backhaus, K., B. Erichson, W. Plinke, R. Weiber (1996) Multivariate Analysemethoden, ed 8. Berlin, Springer.

Balls, M., B.J. Blaauboer, D. Brusick, J. Frazier, D. Lamb, M. Pemberton, C.A. Reinhardt, M. Roberfroid, H. Rosenkranz, B. Schmid, H. Spielmann, A.-L. Stammati, E. Walum (1990) Report and recommendations of the CAAT/ERGATT Workshop on the Validation of Toxicity Test Procedures. ATLA *18:* 313–337.

Balls, M., B.J. Blaauboer, J.H. Fentem, L. Bruner, R.D. Combes, B. Ekwall, R.J. Fielder, A. Guillouzo, R.W. Lewis, D.P. Lovell, C.A. Reinhardt, G. Repetto, D. Sladowski, H. Spielmann, F. Zucco (1995) Practical aspects of the validation of toxicity test procedures. The report and recommendations of ECVAM Workshop 5. ATLA *23:* 129–147.

Bortz, J. (1993) Statistik, ed 3. Berlin, Springer.

Brown, N.A., H. Spielmann, R. Bechter, O.P. Flint, S.J. Freeman, R.J. Jelinek, E. Koch, H. Nau, D.R. Newall, A.K. Palmer, J.-Y. Renault, M. Repetto, R. Vogel, R. Wiger (1995) Screening chemicals for reproductive toxicity: The current alternatives. The report and recommendations of an ECVAM/ETS workshop (ECVAM workshop 12). ATLA *23:* 868–882.

Curren, R.D., J.A. Southee, H. Spielmann, M. Liebsch, J.H. Fentem, M. Balls (1995) The role of prevalidation in the development, validation and acceptance of alternative methods (the first report of the ECVAM Task Force on Prevalidation). ATLA *23:* 211–217.

Doetschman, T., H.R. Eistetter, M. Katz, W. Schmidt, R. Kemler (1985) The in vitro development of blastocyst-derived embryonic stem cell lines: Formation of visceral yolk sac, blood islands and myocardium. J Embryol Exp Morphol *87:* 27–45.

Evans, M.J., M.H. Kaufman (1981) Establishment in culture of pluripotential cells from mouse embryos. Nature *292:* 154–156.

Flint, O.P., T.C. Orton (1984) An in vitro assay for teratogens with cultures of rat embryo midbrain and limb cells. Toxicol Appl Pharmacol *76:* 383–395.

Genschow, E., G. Scholz, N.A. Brown, A.H. Piersma, M. Brady, N. Clemann, H. Huuskonen, F. Paillard, S. Bremer, H. Spielmann (1999) Die Entwicklung von Prädiktionsmodellen für drei in vitro Embryotoxizitätstests in Rahmen einer ECVAM Validierungsstudie. ALTEX *16:* 73–83.

Heuer, J., I.M. Graeber, I. Pohl, H. Spielmann (1994) Culture system for the differentiation of murine embryonic stem cells – A new approach to in vitro testing for embryotoxicity and for developmental immunotoxicology; in Fracchia GN (ed): European Medicines Research. Amsterdam, IOS Press, pp 134–145.

ICH International Conference on Harmonisation of Technical Requirements for Registration of Pharmaceuticals for Human Use (1993) ICH Harmonised Tripartite Guideline: Detection of Toxicity to Human Reproduction for Medicinal Products. Recommended for Adoption at Step 4 of the ICH Process on 24 July 1993 by the ICH Steering Committee. Geneva, ICH Secretariat.

Kochhar, D.M. (1997) Retinoids; in Kavlock, R.J., G.P. Daston (eds): Handbook of Experimental Pharmacology, vol 124/II: Drug Toxicity in Embryonic Development. Berlin, Springer, pp 13–29.

Kolossov E., B.K. Fleischmann, Q. Liu, W. Bloch, S. Viatchenko-Karpinski, O. Manzke, G.J. Ji, H. Bohlen, K. Addicks, J. Hescheler (1998) Functional characteristics of ES cell-derived cardiac precursor cells identified by tissue-specific expression of the green fluorescent protein. J Cell Biol *143:* 2045–2056.

Laschinski, G., R. Vogel, H. Spielmann (1991) Cytotoxity test using blastocyst-derived euploid embryonal stem cells: A new approach to in vitro teratogenesis screening. Reprod Toxicol *5:* 57–64.

Maltsev, V.A., J. Rohwedel, J. Hescheler, A. Wobus (1993) Embryonic stem cells differentiate in vitro into cardiomyocytes representing sinusnodal, atrial and ventricular cell types. Mech Dev *44:* 41–50.

Maltsev, V.A., A.M. Wobus, J. Rohwedel, M. Bader, J. Hescheler (1994) Cardiomyocytes differentiated in vitro from embryonic stem cells developmentally express cardiac-specific genes and ionic currents. Circ Res *75:* 233–244.

Metzger, J.M., W.I. Lin, L.C. Samuelson (1996) Vital staining of cardiac myocytes during embryonic stem cell cardiogenesis in vitro. Circ Res *78:* 547–552.

Newell, D.R., K.E. Beedles (1994) The stem cell test – A novel in vitro assay for teratogenic potential. Toxicol In Vitro *8:* 697–701.

Norušis, M.J. (1994) SPSS Professional Statistics™ 6.1. Chicago, SPSS.

Rohwedel, J., K. Guan, A.M. Wobus (1999) Induction of cellular differentiation by retinoic acid in vitro. Cells Tissues Organs *165:* 190–202.

Rohwedel, J., V.A. Maltsev, E. Bober, H.-H. Arnold, J. Hescheler, A.M. Wobus (1994) Muscle cell differentiation of embryonic stem cells reflects myogenesis in vivo: Developmentally regulated expression of myogenic determination genes and functional expression of ionic currents. Dev Biol *164:* 87–102.

Scholz, G., E. Genschow, I. Pohl, S. Bremer, M. Paparella, H. Raabe, F. Moldenhauer, J. Southee, H. Spielmann (1999) Prevalidation of the embryonic stem cell test (EST), a new in vitro embryotoxicity test. Toxicology In Vitro *13:* 675–681.

Scholz, G., I. Pohl, A. Seiler, S. Bremer, N.A. Brown, A.H. Piersma, H.G.H. Holzhütter, H. Spielmann (1998) Ergebnisse der ersten Phase des ECVAM Projektes zur Prävalidierung und Validierung von drei in vitro Embryotoxizitätstests. ALTEX *15:* 3–8.

Smith, M.K., G.L. Kimmel, D.M. Kochhar, S.P. Shepard, S.P. Spielberg, J.G. Wilson (1983) A selection of candidate compounds for in vitro teratogenesis test validation. Teratogenesis Carcinogenesis Mutagenesis *3:* 461–480.

Southee, J.A., R.D. Curren (1997) A prevalidation process. Comments Toxicol *6:* 27–35.

Spielmann, H. (1998) Reproduction and development. Environ Health Perspect *106:* 571–576.

Spielmann, H., I. Pohl, B. Döring, M. Liebsch, F. Moldenhauer (1997) The embryonic stem cell test (EST), an in vitro embryotoxicity test using two permanent mouse cell lines: 3T3 fibroblasts and embryonic stem cells. In vitro Toxicol *10:* 119–127.

Strübing, C., G. Ahnert-Hilger, S. Jin, B. Wiedenmann, J. Hescheler, A.M. Wobus (1995) Differentiation of pluripotent embryonic stem cells into the neuronal lineage in vitro gives rise to mature inhibitory and excitatory neurons. Mech Dev *53:* 1–13.

Thomas, K.R., M.R. Capecchi (1987) Site-directed mutagenesis by gene targeting in mouse embryo-derived stem cells. Cell *51:* 503–512.

Wiles, M.V. (1993) Embryonic stem cell differentiation in vitro. Methods Enzymol *225:* 900–917.

Wiles, M.V., G. Keller (1991) Multiple hematopoietic lineages develop from embryonic stem (ES) cells in culture. Development *111:* 259–267.

Williams, R.L., D.J. Hilton, S. Pease, T.A. Willson, C.L. Stewart, D.P. Gearing, E.F. Wagner, D. Metcalf, N.A. Nicola, N.M. Gough (1988) Myeloid leukemia inhibitory factor maintains the developmental potential of embryonic stem cells. Nature *336:* 684–687.

Wobus, A.M., G. Wallukat, J. Hescheler (1991) Pluripotent mouse embryonic stem cells are able to differentiate into cardiomyocytes expressing chronotropic responses to adrenergic and cholinergic agents and Ca^{2+} channel blockers. Differentiation *48:* 173–182.

Cells
Tissues
Organs

Cells Tissues Organs 1999;165:212–219

Chicken Embryonic Stem Cells and Transgenic Strategies

B. Pain P. Chenevier J. Samarut

Laboratoire de Biologie, CNRS UMR 49-LA INRA 913, Ecole Normale Supérieure de Lyon, France

Key Words
Transgenesis · Embryonic stem cells, chicken · Transfection · GFP protein

Abstract
The production of transgenic birds is an important goal for both fundamental and applied biology. Different methods have been employed to generate transgenic chickens, including microinjection, use of retroviruses and transfection of primordial germ or embryonic germ cells. In this review we will briefly describe these techniques and our efforts to obtain genetically modified avian embryonic stem (ES) cells using liposomes. This latter technique should allow us to modify chicken ES cells with a high efficiency, permitting the rapid generation of transgenic bird lines.

Abbreviations used in this paper

CES	chicken embryonic stem cells
EG	embryonic germ cells
ES	embryonic stem cells
FBS	fetal bovine serum
PG	primordial germ cells
PI	propidium iodide

Introduction

Embryonic stem (ES) cells were first isolated from the mouse during the 1980s [Evans and Kaufman, 1981]. These cells have unique properties making them receptive to genetic modifications [Bradley et al., 1983; Robertson, 1987]. The relative ease with which to manipulate these cells in vitro have made them a powerful tool to target endogenous genes, and this has led to a dramatic increase in the number of targeted transgenic mice throughout the 1990s. The challenge has been to isolate and characterize cells with similar properties from other species. A number of years passed before ES and/or embryonic germ (EG) cells were isolated from nonmurine, mammalian species like the medaka [Hong et al., 1998], rabbit [Schoonjans et al., 1996], pig [Wheeler, 1994], monkey [Thomson et al., 1995] and human [Thomson et al., 1998], but mammalian germ-line transmission has only been successfully demonstrated in the mouse. With the application of nuclear transfer technology to large farm animals [Campbell et al., 1996; Schnieke et al., 1997; Wilmut et al., 1997; Cibelli et al., 1998], progress towards widespread animal transgenesis appears to be in a revolutionary period.

Other than mouse, chicken is the only species that has shown successful germ-line transmission of transgenes. Modification of the avian genome, although difficult, has important implications for three areas of research. First, the avian model is a preferred developmental model for basic and fundamental research: early embryo develop-

KARGER

Fax + 41 61 306 12 34
E-Mail karger@karger.ch
www.karger.com

1422–6405/99/1654–0212$17.50/0

Accessible online at:
www.karger.com/journals/cto

Dr. B. Pain
Vivalis SA, CHW Nantes – HNB
Place A.-Ricordeau
F–44093 Nantes Cedex 1 (France)
E-Mail bertrandpain@vivalis.com

ment is easy to follow and the embryo itself is easy to manipulate [Dieterlen-Lievre, 1997]. Moreover, comparisons between different species help to define general or specific biological mechanisms. Second, the chicken should be viewed as a powerful bioreactor. The ability to modify protein expression in the egg may prove fruitful in producing molecules with therapeutic interests. Third, the ability to produce poultry resistant to disease or harboring specific traits should be of interest to agronomists [Sang, 1994; Etches and Verrinder Gibbins, 1997]. The aim of this paper is to compare different strategies employed to generate transgenic birds, and we also present data on how genetic modifications of chicken ES cells can be generated efficiently using liposomes.

Methods and Procedures

Cell Culture

Chicken embryonic stem (CES) cells were identified, characterized and employed in vitro and in vivo as previously described by Pain et al. [1996] with slight modifications. Growth factors and cytokines were added as described by Pain et al. [1996]. The transfection medium was similar to the growth medium except that the amount of sera was reduced to 5% for fetal bovine serum (FBS) and 1% for chicken serum.

STO and DT40 cell lines were obtained from ATCC (Rockville, Md., USA). STO cells were maintained in DMEM (BioMedia, France) containing 4.5 g/l glucose, 5% FBS (batch 694-02, Roche, France), 1% *L*-glutamine (BioMedia) and 1% penicillin streptomycin (BioMedia), and DT40 cells cultivated in DMEM (BioMedia) containing 4.5 g/l glucose, 8% tryptose phosphate broth (Difco, USA), 5% FBS (batch 694-02, Roche), 1% chicken serum (Life Technologies, France), 1% penicillin streptomycin (BioMedia) and 0.5 m*M* β-mercaptoethanol (Sigma, France). At confluence, STO cells were dissociated with 0.025% pronase (Roche), washed with PBS containing 5.6 m*M* glucose and plated at 10^4 cells/cm^2 in fresh medium in 100-mm dishes (Costar, USA). At confluence, DT40 cells were centrifuged, washed and replated at a concentration of 10^5 cells/ml.

Transfection Assays

pEGFP-C1 was obtained from Clontech Laboratories (Palo Alto, Calif., USA), prepared by standard procedures using alkaline lysis and PEG purification. 1 µg of circular DNA was used for transient transfection experiments. A variable volume of lipofectant was diluted to a final volume of 100 µl in HamF12 media (BioMedia, France) for FuGENE 6 (Roche, France), and in LipofectAMINE (Life Technologies, France) or Hepes-buffered salt (composed of Hepes 10 m*M*, NaCl 100 m*M*) for DOSPER (Roche) or ExGEN (Eurogentec, Belgium), respectively. 1 µl of the plasmid pEGFP-C1 (1 µg/µl) was added to this mix. The mixture was incubated at room temperature for 15 min and then added dropwise to each well. After 6 h of incubation, the medium was changed and the cells were incubated for a further 36 h. For transfection assays, 5 × 10^4 STO cells or 10^4 DT40 cells were seeded in a 35-mm tissue culture well and cultured overnight in 2 ml of medium. Medium was changed just before transfection. For transfection assays, CES cells were washed with PBS-G, then dissociated with 0.025% pronase (Roche) in citrate buffer, counted, plated on 6-well plates (Costar, France) coated with 0.1% gelatin (Sigma, France) at 5 × 10^5 cells/well in growth medium and cultured overnight. The medium was replaced with 2 ml of fresh transfection medium (see above). After 6 h of incubation, the medium was replaced with fresh growth medium and the cells incubated for 36 h. 2 days after transfection, CES and STO cells were detached using pronase, or just centrifuged for DT40 cells. Cells were washed and resuspended in 1 ml PBS. CES cells were treated with 20 µg propidium iodide (PI). Fluorescence was analyzed by a FACS scan (Becton Dickinson European HQ, Belgium), and calculations were made with the CellQuest software on 10,000 events for STO cells and 5,000 PI-negative events for CES cells.

Results and Discussion

Transgenesis is a unique and powerful tool to modify the genomes of species and its application to the bird has until recently been of limited success. Depending on the stage of development, various approaches could be adopted to introduce genetic modifications into chicken, and multiple strategies have been employed to generate transgenic birds. These techniques are summarized in table 1. The most promising technique is based on work with ES and/or EG cells similar to those initially developed for the mouse [Matsui et al., 1992].

Direct Microinjection of DNA

The first method developed to generate transgenic birds was through direct microinjection of naked DNA into the germinal disc of fertilized ovum. This approach appeared to be particularly difficult to reproduce and not readily applicable on a large scale. The avian ovum is difficult to manipulate because of its 'yolky' structure and masking of the male and female pronuclei by the opaque cytoplasm. Moreover, overly abundant male pronuclei are present [Perry, 1987]. Despite these obstacles, multiple attempts to express transgenes were reported in avian embryos; however, expression in extraembryonic tissues was also reported [Naito et al., 1991a; Perry and Sang, 1993; Naito et al., 1994]. Successful germ-line transmission of a β-galactosidase reporter gene was first demonstrated by Love et al. [1994]. One chimeric male chicken transmitted the DNA transgene to 3.4% of its G1 offspring. The mendelian transmission to the G2 offspring suggested a stable integration event. Even though the approach was successful, the high costs of killing numerous hens to obtain ovocytes and the poor results led to a progressive abandonment of the microinjection tech-

Table 1. Different approaches for avian transgenesis

Developmental stage	Cells	Manipulation	Observations	References
Ovulation	Single cell	Microinjection	Poor efficiency One hen, one ovocyte Not controlled for integration	Love et al. [1994] Naito [1997] Sherman et al. [1998]
Oviposition	Blastodermal cells (CE cells)	Retroviral infection	Not controlled for integration Level of expression	Crittenden et al. [1989] Thoraval et al. [1995]
		Transfection in vitro	In vitro amplification In vitro modification Low germ-line transmission	Petitte et al. [1990] Pain et al. [1996]
Incubation				
Stage 10–11 (33–45 h)	PG cells in germinal crescent	Transfection in ovo	Poor efficiency Not controlled for integration	Kagami et al. [1997]
Stage 14–15 (50–55 h)	PG cells in bloodstream	Transfection in ovo	Poor efficiency Not controlled for integration	Allioli et al. [1994]
		Transfection in vitro	In vitro amplification In vitro modification (?) Germ-line transmission (?)	Kuwana [1993] Tajima et al. [1993] Watanabe et al. [1994]
Stage 27 (5 days)	PG cells in gonads	Transfection in vitro	In vitro amplification In vitro modification (?) Germ-line transmission (?)	Allioli et al. [1997]

nique [Naito, 1997]; however, the recent transposon-based vector used by Sherman et al. [1998] may open new perspectives for this valuable but delicate technique. One advantage of this strategy is the complete development of the chick embryo in vitro based on the original system developed by Perry [1988]. Alternatives to direct microinjection have been developed, including the in ovo microparticle bombardment or the in ova electroporation, the description of which can be seen elsewhere [Muramatsu et al., 1997].

Use of Retroviruses

The use of retroviruses to produce transgenic chickens was a very promising technique because of the efficiency of infectivity. In early studies, viral particles were either directly injected into the embryo or used to infect fresh blastodermal cells in vitro before reimplantation into a recipient embryo. The generation of transgenic birds by retroviral-mediated techniques was first reported by Salter et al. [1987] and Crittenden et al. [1989]; however, the use of replicative competent viruses restricted the use of these animals. The safety of this approach was further questioned because of the presence of endogenous viruses in various strains of chickens and the dissemination of infectious particles. Modification of the viral genome and the use of replication-defective retroviruses overcame many of these concerns [Cosset et al., 1990]. Infection of embryos at the blastodermal stages with defective retroviral particles also produced transgenic chickens [Bosselman et al., 1989; Thoraval et al., 1995]. In these transgenic chickens, no infectious particles could be detected even after 2 generations. The use of retroviruses does not allow a strict control of the resulting genetic modifications. The levels of transgene expression introduced by retroviruses appeared to dwindle, reflecting the viral LTR transcription level. Its expression thus decreased after infection. Another example of retroviral-mediated transgenesis was its use in introducing the β-galactosidase marker into isolated primordial germ (PG) cells [Allioli et al., 1994].

Cellular Approach

In spite of the multiple techniques developed by different laboratories, it appeared that genetic modification might be efficient only through a cellular approach. This approach was facilitated by the isolation of PG cells and the culture of blastodermal cells. Both led to the isolation of cells that can be manipulated in vitro. This strategy was further facilitated by the observation made by Petitte et al. [1990]: cells from a stage X blastoderm [EG and K, Eyal-Giladi and Kochav, 1976] of a pigmented strain grafted to a nonpigmented strain recipient embryo of the same developmental stage produced somatic and germ-line competent chimeras able to exhibit the donor cells' feather color. Progenitor cells responsible for the chimerism were not, however, identified or isolated. Progressively, these grafting experiments were successfully reproduced by other laboratories [Naito et al., 1991b; Watanabe et al., 1992; Thoraval et al., 1994]. Improvement of the technique by slight irradiation of the recipient embryo was also achieved [Petitte et al., 1990; Carsience et al., 1993], and elimination of the cells from the center of the area pellucida of a stage X embryo by replacement of the recipient germ cells with donor germ cells led to a significant increase in donor cell germ-line transmission [Kagami et al., 1997].

Isolation and Manipulation of Circulating PG Cells

Chicken PG cells originate from the ventral surface of the area pellucida in a scattered pattern. They translocate from the epiblast to the hypoblast during early stages of primitive streak formation. They finally concentrate in a region called the 'germinal crescent' located in the anterior extraembryonic area. During all morphological movements, PG cells seem to be passive. They acquire their specific cytological morphology when they are located in the germinal crescent. An extensive number of reports have been devoted to the mechanisms of migration and differentiation of PG cells [Kuwana, 1993; Petitte et al., 1997]; however, few molecular markers have been identified. PG cells finally enter the developing vascular system and circulate temporarily throughout the embryo. This ability to circulate allows their isolation from the bloodstream; therefore, PG cells can be directly isolated and transferred into a recipient embryo. This approach has been successfully employed to produce offspring derived from donor PG cells [Simkiss et al., 1989]. By concentrating the PG cells, a high transmission rate of donor-derived PG cells has been obtained [Yasuda et al., 1992; Tajima et al., 1993; Naito et al., 1994]. Once the cells were isolated and manipulated, a key question was whether isolated PG cells were amenable to genetic modification by transfection [Watanabe et al., 1994; Allioli et al., 1997]. Despite detection of a transgene in the embryo after reimplantation, transmission of the transgene to the offspring could not be demonstrated. Derivation of EG cells from these isolated PG cells might be one way of overcoming these limitations.

Isolation of Avian ES Cells

In nonmammalian species, especially in chicken, the structure of the avian blastoderm differs slightly from those of mammals, complicating the quest for generating avian ES cells. We succeeded in isolating and cultivating chicken ES cells as permanent lines in 1996 [Pain et al., 1996]. Based on the previous observations made by Petitte et al. [1990], we demonstrated that blastodermal cells with ES cell features could be isolated, maintained and amplified in vitro [Pain et al., 1996]. These CES cells were characterized by the presence of different antigens (mainly ECMA-7, EMA-1 and SSEA-1) on their surface, known to be specific to ES cells. We were also the first to detect telomerase activity in these chicken ES cells. Thus the criterion which has recently been used to characterize human ES cell lines [Thomson et al., 1998] appears to be a very sensitive one to discriminate between ES cells and more differentiated cells. A major limitation, however, remains: how is it possible to efficiently introduce genetic modifications into these cells for the generation of transgenic birds?

Genetic Modifications of CES Cells

To generate CES cells with genetic modifications, a number of experiments were performed, taking advantage of prior work on mouse ES cells and commercially available lipofectant reagents. These results generated on STO feeder cells and CES cells are described below.

Determination of Optimal Conditions for Transfection of STO Cells. Tests of transfection efficiencies were performed on STO cells using the enhanced green fluorescence protein as a reporter. We tested several commercial transfection products: ExGEN, LipofectAMINE, DOSPER and FuGENE 6. Each was tested by varying the ratios of lipofectant to DNA. According to the procedures recommended by the companies, the first three compounds were transfected in the absence of serum. LipofectAMINE gave the best results followed by DOSPER and ExGEN, respectively, for a lipofectant to DNA ratio of 10, 8 and 8 µg/µl; however, the transfection efficiencies remained low (<5%). The four lipofectants were then tested in the presence of serum. Although a decrease in transfection efficiency was excepted, the transfection efficiency actually improved for LipofectAMINE and DOSPER. Moreover, the optimal lipofectant to DNA

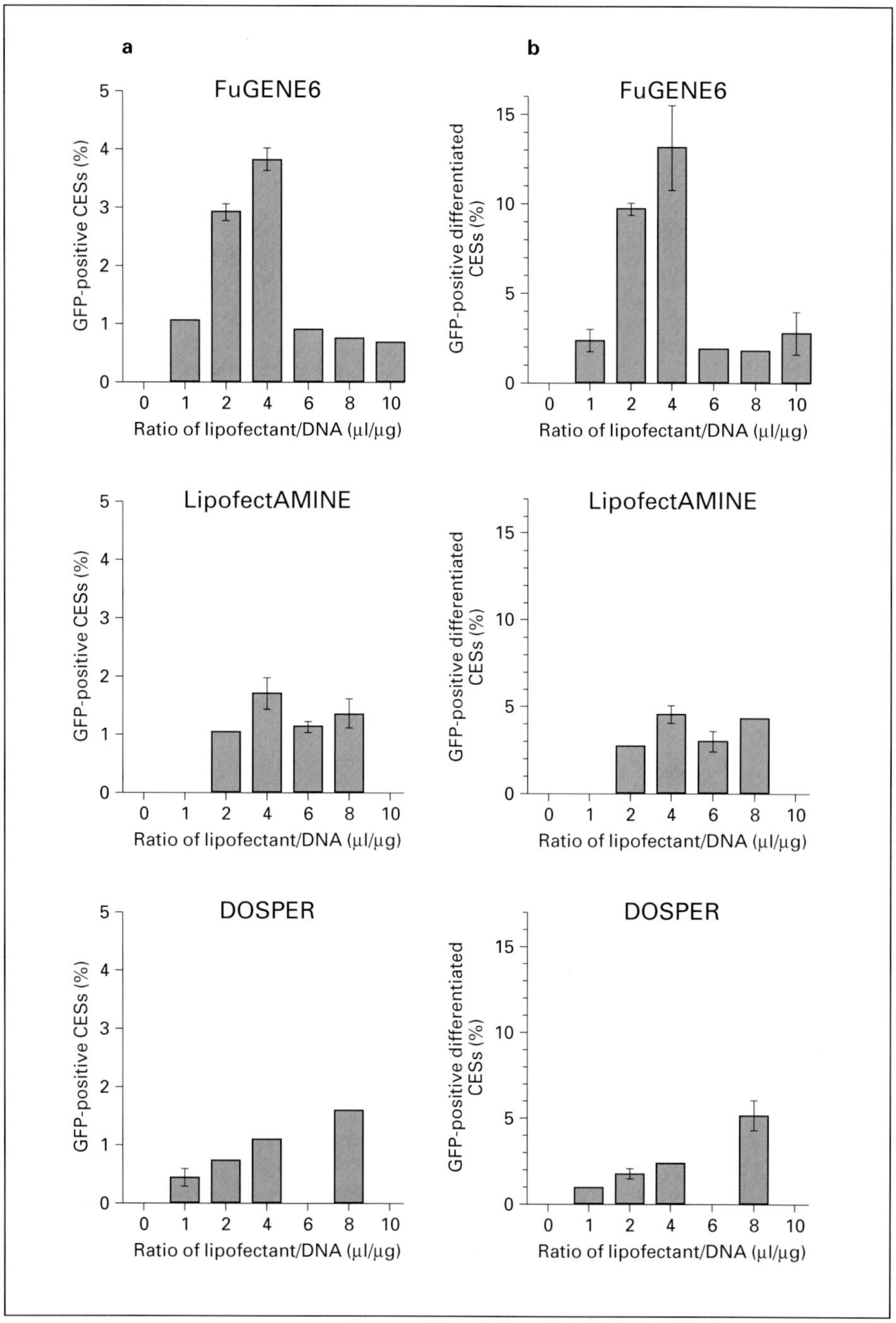

Fig. 1. Comparative efficiency between different liposomes on undifferentiated CES cells (**a**) and differentiated CES cells (**b**). GFP-positive cells were scored by FACS analysis as described in Methods and Procedures.

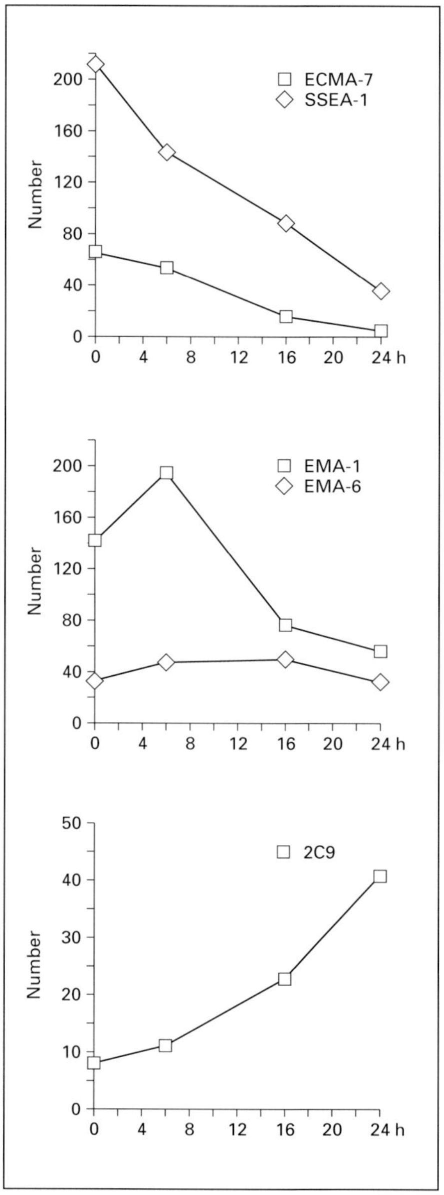

Fig. 2. Number of positive colonies derived per embryo. Embryos were incubated during 0, 6, 16 or 24 h. Dissociated cells were then plated and incubated for 72, 66, 56 and 48 h, respectively. Detection of positive colonies was performed as previously described [Pain et al., 1996].

ratio was found to be lower in the presence of serum (around 3 µg/µg). Although the cytotoxic effects could not be measured easily, lipofection mixtures seemed to be toxic at high lipofectant to DNA ratios. In particular, ExGEN and LipofectAMINE induced a dramatic decrease in cell density at concentrations higher than 10 µg lipofectant/µg DNA. This effect was greatly reduced if the transfection product was used in the presence of serum and was even undetectable for FuGENE 6.

Table 2. Optimal conditions for transfection

	STO cells	DT40 cells	CES cells	Differentiated CES cells
Yield,%	25	<1	4	15
Lipofectant	LipofectAMINE	DOSPER	FuGENE 6	FuGENE 6
Ratio of lipofectant/ DNA, µl/µg	2	>8	4	4

Transfection of CES Cells. We first tried to transfect CES cells with the enhanced green fluorescence protein construct in the absence of serum. CES cells do not survive exposure to serum-free medium, even for short periods of time. We, therefore, obtained low transfection efficiencies. To overcome these problems, transfections were carried out in the presence of serum. During exposure to the lipofection mixture, serum concentrations were reduced to that necessary to cultivate STO cell lines (6 instead of 12%). This permitted comparisons of transfection efficiencies between different cell types. CES cells are fragile and are usually maintained in the presence of inactivated feeder cells to maintain ES cell morphology. In this transfection protocol, we plated them on gelatin for a short period of time, which resulted in rapid morphological changes for some cells. After transfection, CES cell cultures contained two populations: undifferentiated and differentiated cells. Undifferentiated cells are a homogeneous population of small spherical cells; differentiated cells are a morphologically heterogeneous population that includes larger cells. By PI labelling and FACS analysis it was possible to exclude dead cells and distinguish between the two populations by their size and fluorescence characteristics. Estimation of transfection efficiencies for each cell type was thus facilitated (fig. 1). The results demonstrated that differentiated cells are about 3 times more receptive to lipofection than undifferentiated CES cells, whatever the reagent and lipofectant to DNA ratio used. It should be noted that transfected and undifferentiated CES cells have a lower mean fluorescence than transfected differentiated cells (about 1 log lower). FuGENE 6 gave the best transfection efficiencies for a lipofectant to DNA ratio of 4, which is similar to the recommendations of the company. LipofectAMINE and DOSPER proved to be 2-fold less efficient than FuGENE 6. Finally, we could not find any significant correlation between transfection and differentiation rates, but a higher cytotoxicity at high lipofectant to DNA ratio was observed particularly for LipofectAMINE. The comparative results for optimal transfection results are shown in table 2.

Recently, we succeeded in stably modifying CES cells at the genetic level. Two techniques were chosen: one used a test vector containing a CMV promoter and the EGFP reporter as described here and the other involved use of a gene trap vector. The results presented above refer to the conditions of transient transfection and generation of modified cells. As described in the results and in figure 1, transfection of pEGFP-C1 with the FuGENE-6 liposome appeared to provide a relatively high transfection efficiency, and more importantly, the morphology of the transfected cells was well preserved. Selection by G418 of stable CES clones confirmed the success of this transfection procedure (data not shown). These efficient transfection conditions will now allow us to use modified cells for generating chimeric embryos and transgenic chickens.

Relationship between CES and PG Cells

Since CES cells maintained in culture show some spontanous differentiation during transfection (FACS analysis), we investigated the origin of CES cells in the embryo with particular attention to the relationship between PG cells and the CES cells. In order to address this issue, we assayed for the presence of antigens on cells derived from the blastoderm at various times of incubation (no incubation, 6, 16 and 24 h of incubation). We dispersed the cells and maintained them in culture for a maximum of 72 h. We detected the epitopes for ECMA-7, SSEA-1, EMA-1, EMA-6 and 2C9 at the cell surface as previously described [Pain et al., 1996]. Colonies derived from different stages exhibited different reactivities toward these antibodies (fig. 2). Colonies grown in vitro had roughly the same size and nearly the same morphology. Reactivity with antibodies such as ECMA-7 and SSEA-1 was rapidly lost. In contrast the presence of EMA-1- and EMA-6-positive colonies increased. Finally, the presence of 2C9-positive colonies may reflect the emergence of differentiated PG cells. These observations strongly suggest that ECMA-7- and SSEA-1-positive cells are the most immature cell types and that these epitopes are lost during differentiation in vivo. These colonies could produce CES cells. The EMA-

1- and EMA-6-positive cells could reflect the appearance of pre-PG cells, which could give rise to EG-like cells. Finally, cells expressing the 2C9 antibody, which recognized only differentiated PG cells [Maeda et al., 1994], appear only at later stages. Based on antigenic reactivity, these results suggest that fully differentiated PG cells (2C9 positive) only appear 24 h after the emergence of more ECMA-7-positive immature cells.

Conclusions

We have successfully isolated ES-like cells from chicken, and suggest that similar techniques can be applied to isolate ES-like cells from other avian species. Identifying the best conditions to efficiently transfect these cells will be the key to the ultimate success in generating transgenic birds. Further work is however needed to determine their contribution to somatic cells and to evaluate the germ-line competence of these cells which is currently very low.

References

Allioli, N, J.L. Thomas, Y. Chebloune, V.M. Nigon, G. Verdier, C. Legras (1994) Use of retroviral vectors to introduce and express the β-galactosidase marker gene in cultured chicken primordial germ cells. Dev Biol *165:* 30–37.

Allioli, N., G. Verdier, C. Legras (1997) Use of gonadal primordial germ cells (PGCs) as tools for gene transfer in chickens. Methods Mol Biol *62:* 425–432.

Bosselman, R.A., R.Y. Hsu, T. Boggs, S. Hu, J. Bruszewski, S. Ou, L. Kozar, F. Martin, C. Green, F. Jacobsen, M. Nicholson, J.A. Schultz, K.M. Semon, W. Rishell, R.G. Stewart (1989) Germline transmission of exogenous genes in the chicken. Science *243:* 533–535.

Bradley, A., M. Evans, M. Kaufman, E.J. Robertson (1984) Formation of germ line chimeras from embryo-derived teratocarcinoma cell lines. Nature *309:* 255–256.

Campbell, K.H., J. McWhir, W.A. Ritchie, I. Wilmut (1996) Sheep cloned by nuclear transfer from a cultured cell line. Nature *380:* 64–66.

Carsience, R.S., M.E. Clark, A.M. Verrinder Gibbins, R.J. Etches (1993) Germline chimeric chickens from dispersed donor blastodermal cells and compromised recipient embryos. Development *177:* 669–675.

Cibelli, J.B., S.L. Stice, P.J. Golueke, J.J. Kane, J. Jerry, C. Blackwell, F.A. Ponce de Leon, J.M. Robl (1998) Transgenic bovine chimeric offspring produced from somatic cell derived stem-like cells. Nat Biotechnol *16:* 642–646.

Cosset, F.L., C. Legras, Y. Chebloune, P. Savatier, P. Thoraval, J. Thomas, J. Samarut, V.M. Nigon, G. Verdier (1990) A new avian leukosis virus-based packaging cell line that uses two separate transcomplementing helper genomes. J Virol *64:* 1070–1078.

Crittenden, L.B., D.W. Salter, M.J. Federspiel (1989) Segregation, viral phenotype and proviral structure of 23 avian leukosis virus inserts in the germ line of chickens. Theor Appl Genet *77:* 505–515.

Dieterlen-Lievre, F. (1997) Avian models in developmental biology. Poult Sci *76:* 78–82.

Etches, R.J., A.M. Verrinder Gibbins (1997) Strategies for the production of transgenic chicken. Methods Mol Biol *62:* 433–450.

Evans, M.J., M.H. Kaufman (1981) Establishment in culture of pluripotential cells from mouse embryo. Nature *292:* 145–156.

Eyal-Giladi, H., S. Kochav (1976) From cleavage to primitive streak formation: A complementary normal table and a new look at the first stages of the development of the chick. I. General morphology. Dev Biol *49:* 321–337.

Hong, Y., C. Winkler, M. Schartl (1998) Production of medakafish chimeras from a stable embryonic stem cell line. Proc Natl Acad Sci USA *95:* 3679–3684.

Kagami, H., T. Tagami, Y. Matsubara, T. Harumi, H. Hanada, K. Maruyama, M. Sakurai, T. Kuwana, M. Naito (1997) The developmental origin of primordial germ cells and the transmission of the donor derived gametes in mixed-sex chimeras to the offspring in the chicken. Mol Reprod Dev *48:* 501–510.

Kuwana, T. (1993) Migration of avian primordial germ cells torward the gonadal anlage. Growth Differ *35:* 237–243.

Love, J., C. Gribbin, C. Mather, H. Sang (1994) Transgenic birds by DNA microinjection. Biotechnology *12:* 60–63.

Maeda, S., S. Ohsako, M. Kurohmaru, Y. Hayashi, T. Nishida (1994) Analysis for the stage specific antigen of the primordial germ cells in the chick embryo. J Vet Med Sci *56:* 315–320.

Matsui, Y., K. Szebo, B. Hogan (1992) Derivation of pluripotential embryonic stem cells from murine primordial germ cells in culture. Cell *70:* 841–847.

Muramatsu, T., Y. Mizutani, Y. Ohmori, J.I. Okumura (1997) Comparison of three nonviral transfection methods for foreign gene expression in early chicken embryos in ovo. Biochem Biophys Res Commun *230:* 376–380.

Naito, M. (1997) The microinjection of DNA into early chicken embryo; in Houdebine, L.M. (ed): Transgenic Animals: Generation and Use. Hardwood, pp 69–73.

Naito, M., K. Agata, K. Otsuka, K. Kino, M. Ohta, K. Hirose, M.M. Perry, G. Eguchi (1991a) Embryonic expression of β-actin-lacZ hybrid gene injected into the fertilized ovum of the domestic fowl. Int J Dev Biol *35:* 69–75.

Naito, M., E. Sasaki, M. Ohtaki, M. Sakurai (1994) Introduction of exogenous DNA into somatic and germ cells of chicken by microinjection into the germinal disc of fertilized ova. Mol Reprod Dev *37:* 167–171.

Naito, M., M. Watanabe, M. Kinutani, K. Nirasawa, T. Oishi (1991b) Production of quail-chick chimeras by blastoderm cell transfer. Br Poult Sci *32:* 79–86.

Pain, B., M.E. Clark, M. Shen, H. Nakazawa, M. Sakurai, J. Samarut, R.J. Etches (1996) Long-term in vitro culture and characterisation of avian embryonic stem cells with multiple morphogenetic potentialities. Development *122:* 2339–2348.

Perry, M.M. (1987) Nuclear events from fertilisation to the early cleavage stages on the domestic fowl *(Gallus domesticus).* J Anat *150:* 99–109.

Perry, M.M. (1988) A complete culture system for the chick embryo. Nature *331:* 70–72.

Perry, M.M., H.M. Sang (1993) Transgenesis in chickens. Transgenic Res *2:* 125–133.

Petitte, J., M.E. Clark, A.M. Verrinder Gibbins, R.J. Etches (1990) Production of somatic and germline chimeras in the chicken by transfer of early blastodermal cells. Development *108:* 185–189.

Petitte, J.N., L. Karagenc, M. Ginsburg (1997) The origin of the avian germ line and transgenesis in birds. Poult Sci *76:* 1084–1092.

Robertson, E.J. (1987) Embryo derived stem cell lines; in Robertson, E. (ed): Teratocarcinomas and Embryonic Stem Cells: A Practical Approach. Oxford, IRL Press, pp 71–112.

Salter, D.W., E.J. Smith, S. Hughes, S.E. Wright, L.B. Crittenden (1987) Transgenic chickens: insertion of retroviral genes into the chicken germ line. Virology *157:* 236–240.

Sang, H. (1994) Transgenic chickens: Methods and potential applications. Trends Biotechnol *12:* 415–420.

Sang, H.M., M.M. Perry (1989) Episomal replication of cloned DNA injected into the fertilised ovum of the hen *(Gallus domesticus).* Mol Reprod Dev *1:* 98–106.

Schnieke, A.E., A.J. Kind, W.A. Ritchie, K. Mycock, A.R. Scott, M. Ritchie, I. Wilmut, A. Colman, K.H. Campbell (1997) Human factor IX transgenic sheep produced by transfer of nuclei from transfected fetal fibroblasts. Science *278:* 2130–2133.

Schoonjans, L., G.M. Albright, J.L. Li, D. Collen, R.W. Moreadith (1996) Pluripotential rabbit embryonic stem cells are capable of forming overt coat color chimeras following injection into blastocysts. Mol Reprod Dev *45:* 439–443.

Sherman, A., A. Dawson, C. Mather, H. Gilhooley, Y. Li, R. Mitchell, D. Finnegan, H. Sang (1998) Transposition of the Drosophila element mariner into the chicken germ line. Nat Biotechnol *16:* 1050–1053.

Simkiss, K., K. Rowlett, N. Bumstead, B.M. Freeman (1989) Transfer or primordial germ cells DNA between embryos. Protoplasma *151:* 164–166.

Tajima, A., M. Naito, Y. Yasuda, T. Kuwana (1993) Production of germ line chimera by transfer of primordial germ cells in domestic chicken *(Gallus domesticus).* Theriogenology *40:* 509–519.

Thomson, J.A., J. Itskovitz-Eldor, S.S. Shapiro, M.A. Waknitz, J.J. Swiergiel, V.S. Marshall, J.M. Jones (1998) Embryonic stem cell lines derived from human blastocysts. Science *282:* 1145–1147.

Thomson, J.A., J. Kalishman, T.G. Golos, M. Durning, C.P. Harris, R.A. Becker, J.P. Hearn (1995) Isolation of a primate embryonic stem cell line. Proc Natl Acad Sci USA *92:* 7844–7848.

Thoraval, P., M. Afanassieff, F.L. Cosset, F. Lasserre, G. Verdier, F. Coudert, G. Dambrinne (1995) Germline transmission of exogenous genes in chickens using helper free ecotropic avian leukosis virus-based vectors. Transgenic Res *4:* 369–376.

Thoraval, P., F. Lasserre, F. Coudert, G. Dambrinne (1994) Somatic and germline chicken chimeras obtained from brown and white leghorns by transfer of early blastodermal cells. Poult Sci *73:* 1897–1905.

Watanabe, M., M. Kinutani, M. Naito, O. Ochi, Y. Takashima (1992) Distribution analysis of transferred donor cells in avian blastodermal chimeras. Development *114:* 331–338.

Watanabe, M., M. Naito, E. Sasaki, M. Sakurai, T. Kuwana, T. Oishi (1994) Liposome mediated DNA transfer into chicken primordial germ cells in vivo. Mol Reprod Dev *38:* 268–274.

Wheeler, M.B. (1994) Development and validation of swine embryonic stem cells: A review. Reprod Fertil Dev *6:* 563–668.

Wilmut, I., A.E. Schnieke, J. McWhir, A.J. Kind, K.H. Campbell (1997) Viable offspring derived from fetal and adult mammalian cells. Nature *385:* 810–813.

Yasuda, Y., A. Tajima, T. Fujimoto, T. Kuwana (1992) A method to obtain avian germ line chimeras using isolated primordial germ cells. J Reprod Fertil *96:* 521–528.

Cells Tissues Organs

Cells Tissues Organs 1999;165:220–236

Accepted after revision: June 10, 1999

Establishment of Pluripotent Cell Lines from Vertebrate Species – Present Status and Future Prospects

Katja Prelle[a] Ivan M. Vassiliev[b,1] Svetlana G. Vassilieva[b] Eckhard Wolf[a] Anna M. Wobus[b]

[a]Department of Molecular Animal Breeding and Genetics, Gene Centre, Ludwig Maximilian University, Munich and [b]In Vitro Differentiation Group, IPK Gatersleben, Germany

Key Words

Stem cells, embryonic · Primordial germ cells · Domestic animals · Gene transfer · Nuclear transfer

Abstract

Pluripotent embryonic stem (ES) cells are undifferentiated cell lines derived from early embryos and are capable of unlimited undifferentiated proliferation in vitro. They retain the ability to differentiate into all cell types including germ cells in chimeric animals in vivo, and can be induced to form derivatives of all three germ layers in vitro. Mouse ES cells represent one of the most important tools in genetic research. Major applications include the targeted mutation of specific genes by homologous recombination and the discovery of new genes by gene trap strategies. These applications would be of high interest for other model organisms and also for livestock species. However, in spite of tremendous research activities, no proven ES cells colonizing the germ line have been established for vertebrate species other than mouse and chicken thus far. This review summarizes the current status of deriving pluripotent embryonic stem cell lines from vertebrates and recent developments in nuclear transfer technology, which may provide an alternative tool for genetic modification of livestock animals.

K.P. and I.M.V. contributed equally to this work.

[1] *Present address:* Institute of Physiological Chemistry, Martin Luther University, Halle/Saale (Germany).

Abbreviations used in this paper

AP	alkaline phosphatase
bFGF	basic fibroblast growth factor
BRL	Buffalo rat liver
EG	embryonic germ cells
ES	embryonic stem cells
ICM	inner cell mass
IL	interleukin
LIF	leukemia inhibitory factor
neo	neomycin
PGC	primordial germ cells
SCF	stem cell factor
SSEA	stage-specific embryonic antigen

KARGER

Fax + 41 61 306 12 34
E-Mail karger@karger.ch
www.karger.com

1422–6405/99/1654–0220$17.50/0

Accessible online at:
www.karger.com/journals/cto

Dr. Katja Prelle
Department of Molecular Animal Breeding and Genetics
Ludwig Maximilian University Munich, Hackerstrasse 27
D–85764 Oberschleissheim (Germany)
Tel. +49 89 31573233, Fax +49 89 3152799, E-Mail k.prelle@gen.vetmed.uni-muenchen.de

Table 1. Toti-/pluripotency of cells and nuclei [according to Campbell and Wilmut, 1997]

	Cells		Nuclei	
	totipotency	pluripotency	totipotency	pluripotency
Definition	Ability to form a whole organism	Ability to differentiate into many or all tissues including the germ line of chimeric animals	Ability to support full-term development of an embryo reconstructed by nuclear transfer	Ability to partially support the development of nuclear transfer embryos
Examples	Blastomeres of early cleavage stage embryos[1]	ICM cells, EC cells, ES cells, EG cells	Sheep: cultured embryonic disc cells, fetal fibroblasts, mammary gland cells Cattle: fetal fibroblasts, germ cells, postnatal skin fibroblasts, uterine cells Mouse: cumulus cells [reviewed in Wolf et al., 1998]	Cattle: oogonia, trophoblastic cells Mouse: Sertoli cells, neurons [reviewed in Wolf et al., 1998]
Technology	Embryo splitting, blastomere separation	Aggregation with morulae, injection into blastocysts	Nuclear transfer	Nuclear transfer

EC = Embryonic carcinoma cells.
1 Germ line-competent ES and EG cells were also classified as totipotent by some authors.

Pluripotent Cell Lines in Mouse

Cells capable of forming a whole conceptus are termed 'totipotent', whereas 'pluripotent' cells have the ability to contribute to several tissues of the fetus and in some cases to the extraembryonic membranes when used for chimera formation [reviewed in Campbell and Wilmut, 1997]. The developmental potential of cells has to be separated from that of nuclei, which may be derived from differentiated cells but reprogrammed by nuclear transfer and thus give rise to a whole organism (table 1).

In mice, pluripotent embryonic cells were first isolated from teratocarcinomas, which comprise a mixture of differentiated cell types including derivatives of all germ layers and remaining undifferentiated (malignant) stem cells [Martin and Evans, 1975]. These embryonic carcinoma cells show several disadvantages, like overt karyotypic abnormalities and low germ line colonization due to the deregulated environment within teratocarcinomas.

Embryonic stem (ES) cells may be derived from 8-cell embryos [Wobus et al., 1991; Delhaise et al., 1996] or dissociated blastomeres of morulae [Eistetter, 1989], but are most frequently established from the inner cell mass (ICM) of blastocysts [Evans and Kaufman, 1981; Martin, 1981] or the pluripotent cell population of the primitive ectoderm of delayed implanting blastocysts [Brook and Gardner, 1997]. Most ES cell lines were isolated according to standard protocols [Robertson, 1997] from embryos of the inbred strain 129Sv, although selection strategies have been developed to allow isolation of ES cells from embryos with a normally nonpermissive genetic background [McWhir et al., 1996]. Undifferentiated proliferation of ES cells in culture was initially achieved by using feeder cells [primary mouse embryonic fibroblasts, Wobus et al., 1984; STO cells, Evans and Kaufman, 1981; Robertson, 1987] or media conditioned by Buffalo rat liver (BRL) cells [Smith and Hooper, 1987]. The differentiation-inhibiting activity in both systems has been identified as leukemia inhibitory factor (LIF) [Smith et al., 1988]. The recombinant protein maintains pluripotency of ES cells in the absence of feeder cells [Nichols et al., 1990]. Alternatively, other cytokines acting through the gp130 signal transduction pathway, including interleukin-6 (IL-6), oncostatin M and ciliary neurotrophic factor [Conover et al., 1993; Nichols et al., 1994; Rose and Bruce, 1994; Wolf et al., 1994; reviewed in Wolf et al., 1996], have successfully been used to sustain undifferentiated murine ES cells capable of differentiating in vitro and in vivo.

A third class of pluripotent cells in mice is derived from primordial germ cells (PGCs) which migrate from the base of the allantois through the hindgut epithelium and the dorsal mesentery into the gonadal anlage [Eddy et al., 1981]. After a period of intragonadal mitotic proliferation the male germ cells enter a quiescent period of mitotic arrest in the testis, while the female PGCs undergo rapid mitotic divisions before entering the meiotic arrest in the ovary [reviewed in Wylie, 1993]. Embryonic germ (EG) cells can be derived from pre- and postmigratory PGCs by using a combination of LIF, basic fibroblast growth factor (bFGF) and stem cell factor (SCF) in the culture medium [Matsui et al., 1992; Resnick et al., 1992; Stewart et al., 1994] and behave in a similar way as ES cells when injected into host blastocysts [Matsui et al., 1992] and upon induced differentiation in vitro [Rohwedel et al., 1996]. PGCs are available in larger numbers per fetus, but the extent of methylation of specific genes may differ compared to ICM cells [Solter, 1998].

Characteristics of Pluripotent Embryonic Cell Lines in Mouse

Pluripotent cells exhibit morphological characteristics, i.e. a high nuclear:cytoplasm ratio and growth in compact, multilayered colonies. They show alkaline phosphatase (AP) activity [Resnick et al., 1992] and express highly conserved epitope markers, like stage-specific embryonic antigen-1 (SSEA-1) [Solter and Knowles, 1978] or ECMA-7 [Pease et al., 1990], and the germ line-specific transcription factor Oct-4 [Schöler et al., 1989]. In addition, a short G1 cell cycle phase [Rohwedel et al., 1996] and high levels of telomerase activity [Thomson et al., 1998] are highly correlated with the undifferentiated state of ES cells.

Pluripotent cells can be stimulated to differentiate in vitro into various cell types by removing feeder cells and differentiation-inhibiting cytokines [reviewed in Pedersen, 1994], or by adding specific differentiation-inducing substances [e.g. retinoic acid, Wobus et al., 1994, 1997a; see Rohwedel et al., 1999]. Maintained in suspension culture or 'hanging drops' [Wobus et al., 1984], pluripotent cells form embryo-like aggregates, so-called 'embryoid bodies', consisting of derivatives of the three embryonic germ layers [Doetschman et al., 1985; Rohwedel et al., 1996, reviewed in Wobus and Guan, 1998].

In vivo differentiation of pluripotent cells occurs after transplantation to syngenic or immunodeficient mice, where they form teratocarcinomas or teratomas [Evans and Kaufman, 1981]. When combined with normal embryos via blastocyst injection [Bradley et al., 1984] or morula aggregation [Wood et al., 1993], these cells contribute to all tissues and organs including the germ line of a chimeric individual, giving rise to functional gametes [Labosky et al., 1994]. When 'target' ES cells were aggregated with 'carrier' tetraploid embryos [Nagy et al., 1990], the pluripotent cells formed the adult organism without any participation of tetraploid host embryo cells [Wang et al., 1997].

Pluripotent Embryonic Cell Lines in Other Vertebrate Species

Murine pluripotent embryo-derived cell lines offer a suitable model system for investigating cellular differentiation of the vertebrate organism [Wobus et al., 1997b] and provide an important system for site-specific introduction of genetic alterations by gene targeting technology [Thomas and Capecchi, 1987] as well as for screening developmental genes by gene trap experiments [reviewed in Evans et al., 1997]. Because of these powerful applications, many attempts have been made to establish ES and EG cell technology in other species. Although many of these embryo-derived cell lines resemble – at least in part – mouse ES cell morphology and epitope profiles [Van Stekelenburg-Hamers et al., 1995; Etches et al., 1996] and have the ability to differentiate in vitro, there is no published data reporting liveborn, fertile germ line chimeras in mammalian species other than mouse (table 2). Establishment of pluripotent cell lines especially from mammals other than mice would have a great impact on studies of early developmental processes, such as cell lineage commitment [Beddington and Robertson, 1989] and genomic imprinting [Mann et al., 1990], and especially on genetic modification [Wheeler and Choi, 1997] and nuclear transfer in livestock species [Modlinski et al., 1996].

Fish

Fish are excellent vertebrate model systems, especially for research fields of developmental biology, i.e. the molecular analysis of normal and disturbed development.

In zebrafish, blastula-derived embryonic cell lines have been maintained undifferentiated for more than 40 passages using trout embryo extract and serum added to BRL cell-conditioned medium [Collodi et al., 1992]. When using neomycin (neo)-transfected cells, the foreign neo-specific DNA sequences were detected in juvenile

Table 2. Establishment of pluripotent embryonic cell lines from vertebrate species (except mouse)

Species	Cell source	ES-like characteristics	In vivo differentiation	References
Medakafish	Midblastulae	AP activity, EB formation	Transgenic chimeric fry	Hong et al. [1998]
Zebrafish	Midblastulae	AP activity, in vitro differentiation	Transgenic chimeric juveniles	Sun et al. [1995]
Chicken	Stage X blastoderm	AP activity, SSEA-1, SSEA-3 and ECMA-7 expression	Germ line chimeras	Pain et al. [1996]
	5-day genital ridges	Morphology	Germ line chimeric chicken	Chang et al. [1997]
Rabbit	5-day intact embryos	EB formation	Coat color chimeras	Schoonjans et al. [1996]
	18- to 22-day genital ridges	SSEA-1 expression, AP activity	Coat color chimeras	Moens et al. [1997]
Rat (?)	Blastocysts	AP activity, SSEA-1 expression, EB formation	Coat color chimeras (not proven in independent repeated experiments)	Iannaccone et al. [1994]
Syrian hamster	3-day blastocysts	EB formation	n.d.	Doetschman et al. [1988]
Mink	7-day blastocysts	EB formation, teratoma formation	n.d.	Sukoyan et al. [1993]
Pig	Blastocysts	EB formation	Coat color chimeras	Wheeler [1994]
	25- to 27-day genital ridges	AP activity, EB formation	Transgenic chimeric piglet	Piedrahita et al. [1998]
Cattle	Blastocysts derived by fibroblast cloning	Morphology	Transgenic chimeric calves	Cibelli et al. [1998a]
	45-day genital ridges	Morphology, pseudopodia	Cloned bull calf	Strelchenko et al. [1998]
Sheep	8-day blastocysts	Morphology	Cloned lambs	Wells et al. [1997]
Rhesus monkey	6-day blastocysts	AP activity, SSEA-3 and SSEA-4 expression, in vitro differentiation, EB and teratoma formation	n.d.	Thomson et al. [1995]
Common marmoset	8-day blastocysts		n.d.	Thomson et al. [1996]
Human	IVF blastocysts	Morphology, telomerase activity, AP activity, SSEA-3 and SSEA-4 expression	n.d.	Thomson et al. [1998]
	5- to 9-week genital ridges	AP activity, SSEA-1, SSEA-3 and SSEA-4 expression, EB formation	n.d.	Shamblott et al. [1998]

EB = Embryoid body; n.d. = not done.

fish 18 days following blastocyst injection of the ES-like cells. For a detailed analysis embryonic cells from blastula stage embryos were cultured on zebrafish embryonic fibroblasts in medium supplemented with bovine insulin, LIF, an extract of rainbow trout embryos, and trout serum [Sun et al., 1995]. The cells possessed a diploid karyotype, exhibited an ES-like morphology and high AP activity. Induced in vitro differentiation resulted in neuronal cell types. Genetically marked embryonic cells expressed the *neo* gene in chimeric fish generated by blastocyst injection.

Blastomeres isolated from medakafish embryos at the midblastula stage were cultured on homologous fibroblasts with addition of carp serum, LIF and bFGF, and were maintained in culture for up to 11 passages. These cells exhibited morphological features comparable to mouse ES cells with high AP activity [Wakamatsu et al., 1994]. Differentiated derivatives induced by retinoic acid

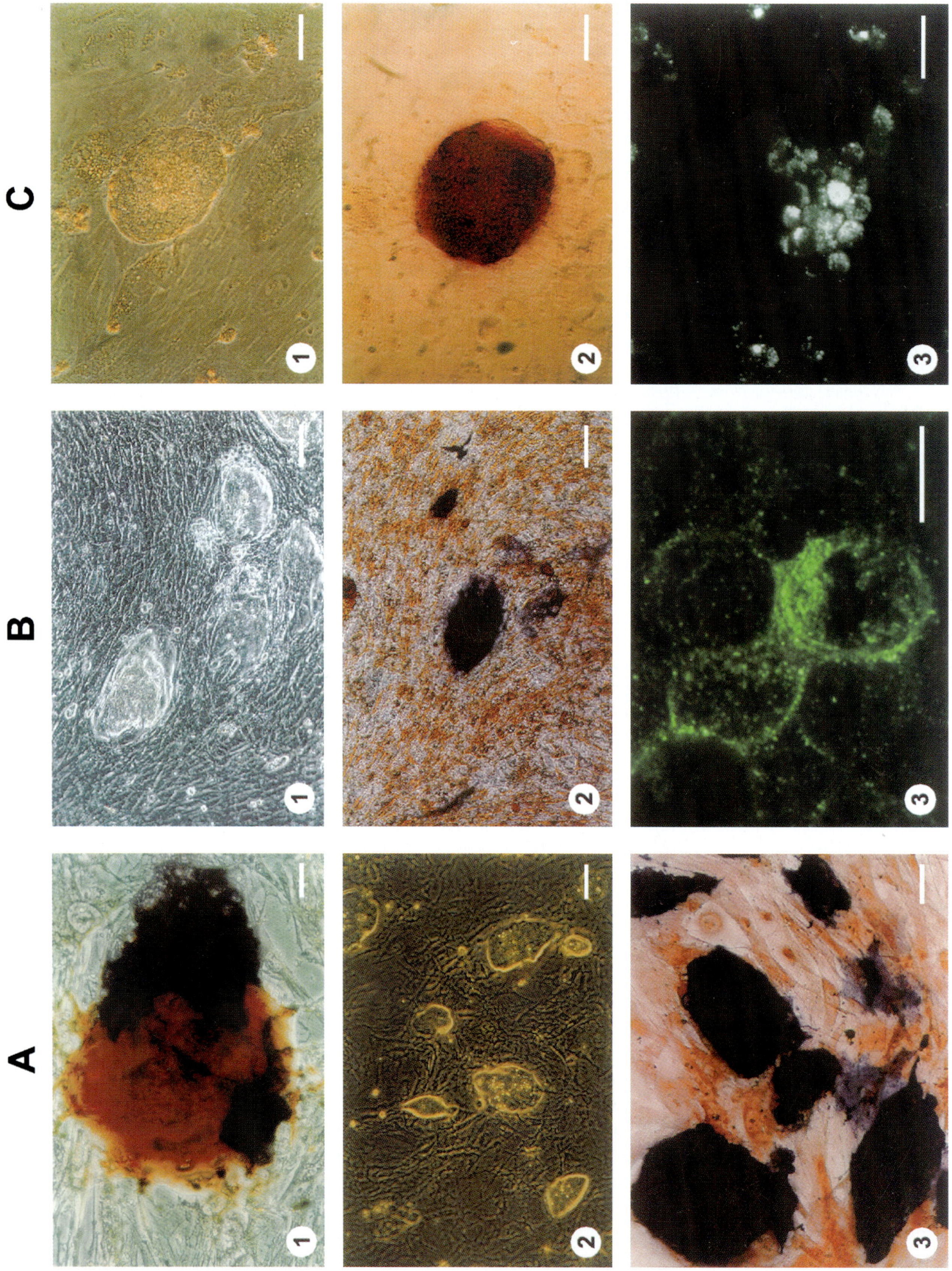
A
B
C
1
2
3
1
2
3
1
2
3

included melanocytes, neuronal and skeletal muscle-like cells.

Hong and Schartl [1996] set up conditions supporting feeder cell-free cultivation of medakafish embryonic cells isolated at the midblastula stage, which displayed features resembling mouse ES cells when cultured in medium supplemented with human bFGF and LIF, fish embryo extract, and trout serum. Continuously cultured over 18 months (100 passages), the putative medakafish ES cells formed densely packed, AP-positive colonies. In suspension culture they developed embryoid bodies and differentiated into various cell types including neuron- and fibroblast-like cells, pigment and muscle cells [Hong et al., 1996]. These ES cells were competent to form chimeras with a high contribution of the transplanted cells to numerous organs [Hong et al., 1998].

Fig. 1. Establishment of pluripotent embryonic cells in culture from rabbit, rat and pig. **A** Rabbit embryos (Zika® albino) as source of pluripotent ES-like cells. **A1** Blastocyst outgrowth with AP-positive ICM counterstained with neutral red (ALP kit, Sigma) at the 1 day after attachment. Bar = 50 µm. **A2** ES-like colonies at passage 7 cultured on rabbit embryonic fibroblasts in Dulbecco's modified Eagle's medium (DMEM)/Ham's F-12 (50:50, Sigma), supplemented with 15% fetal calf serum (FCS; Gibco BRL), 2 m*M* *L*-glutamine (Hyclone), 0.1 m*M* nonessential amino acids (NEAA; Sigma), penicillin/streptomycin (100 IU/100 mg; Gibco BRL), 1% nucleoside stock solution (Sigma) and 1,000 IU/ml murine LIF (ESGRO, Gibco BRL). Bar = 100 µm. **A3** AP-positive rabbit ES-like colonies on embryonic fibroblasts after 17 passages. Bar = 100 µm. **B** Rat ES-like cells established from 4.5-day blastocysts (F-344), cultured on mouse embryonic fibroblasts in DMEM/Iscove's medium (50:50; Gibco BRL), supplemented with 15% FCS, NEAA (1% stock solution; Gibco BRL), 2 m*M* *L*-glutamine (Gibco BRL), 0.3 m*M* each of adenosine, cytidine, guanosine and uridine (Sigma), 0.1 m*M* thymidine (Sigma), human transferrin (15 µg/ml; Boehringer, Mannheim), penicillin/streptomycin (50 IU/50 µg; Gibco BRL), 20 ng/ml human LIF (prepared using plasmid pGEX2T.LIF-58), 450 µM monothioglycerol (Sigma), and 20 ng/ml bFGF (Strathmann Biotech). **B1** Compact colonies at passage 6. Bar = 100 µm. **B2** Histochemical staining for AP activity (ALP kit, Sigma). Bar = 100 µm. **B3** SSEA-1 immunolabelling of rat ES-like cells (RES-6) at passage 6 after confocal laser scanning microscopy. Bar = 10 µm. **C** Porcine PGCs isolated from 27-day-old fetuses, cultured on STO feeder cells in DMEM (Gibco BRL), supplemented with 10% FCS (Boehringer, Mannheim), 1% NEAA (Gibco BRL), 2 m*M* *L*-glutamine (Gibco BRL), penicillin/streptomycin (10 IU/ 100 µg; Gibco BRL). **C1** Multi-layered compact colonies after 28 weeks (15 passages) in vitro. Bar = 50 µm. **C2** AP staining of undifferentiated porcine PGCs after 28 weeks (15 passages; ALP kit, Sigma). Bar = 50 µm. **C3** SSEA-1 immunolabelling of porcine EG cells after 30 days in culture by confocal laser scanning microscopy. Bar = 25 µm.

Avian

In the chicken, embryos at the designated stage X are composed of 40,000–80,000 morphologically undifferentiated cells [Thorval et al., 1994]. These cells, when injected into the subgerminal cavity of host embryos, were incorporated into the germ line and could subsequently contribute to melanocytes and erythrocytes [Petitte et al., 1990]. The somatic and germ line chimerism was increased when the development of recipient embryos was compromised by exposure to γ-irradiation [Carsience et al., 1993]. To propagate these putative pluripotent blastodermal cells in vitro, they were cultured for a short time (48 h) as explanted embryos, as a dispersed cell monolayer or on a mouse fibroblast feeder layer in the presence of LIF, bFGF and chicken SCF [Etches et al., 1996]. Regardless of the culture conditions, the contribution to ectoderm-derived tissues and to the germ line of chimeras produced by injection into the subgerminal cavity was reduced as compared to freshly isolated blastodermal cells. Putative avian ES cells could be maintained long-term in vitro on STO cells in medium supplemented with LIF, bFGF, SCF, oncostatin M, IL-11 and IL-6 [Pain et al., 1996]. These cells exhibited features similar to murine ES cells such as typical morphology, high AP activity, and specific epitopes such as SSEA-1, SSEA-3, ECMA-7 and EMA-1. They presented high capacities of in vitro differentiation, and avian early passage ES cells injected to host embryos were able to contribute extensively to chimeric chickens, but showed low germ line transmission [see Pain et al., 1999].

Avian PGCs segregate from the hypoblast of the germinal crescent region and begin to circulate in the embryonic bloodstream, from which they can be isolated before leaving the capillaries near the germinal epithelium to penetrate the developing gonads [Kuwana et al., 1996]. PGCs isolated from 5.5-day-old chicken embryonic germinal ridges were cultured for 5 days in medium enriched with insulin-like growth factor-1, bFGF and LIF, and contributed to the germ line of chimeric chicken following transfer into the bloodstream of host embryos [Chang et al., 1997].

Rabbit

Rabbits have several advantages over other model organisms for studies of vascular diseases, cardiomyopathy, central nervous system disorders, and abnormalities of lipoprotein metabolism. Other criteria, such as ease in housing, ability to superovulate, manipulation of embryos, and litter size make the rabbit an attractive experimental system [Graves and Moreadith, 1993]. Rabbit

ICM cells were isolated from day 3 embryos by immunosurgery [Giles et al., 1993], or by mechanically disrupting outgrowths of day 4 to 5 blastocysts [Graves and Moreadith, 1993]. As detected by isozyme analysis and eye pigmentation rabbit ICM cells cultured on STO cells contributed to chimeric fetuses following blastocyst injection before they showed decreased mitotic activity resulting in complete differentiation after 5 passages [Giles et al., 1993]. Murine LIF preserved undifferentiated ES-like features including AP activity (fig. 1, A3) and the ability to form embryoid bodies [Graves and Moreadith, 1993; Vasilieva et al., 1998]. Following injection into blastocysts, putative rabbit ES cells contributed to fertile overt coat color chimeras, but no germ line transmission was shown [Schoonjans et al., 1996]. However, nuclear transfer resulted in development to blastocysts at a similar rate as obtained with embryonic blastomeres as nuclear donors [Du et al., 1995].

Rabbit PGCs were isolated from gonadal cells of 18- to 22-day-old fetuses, and displayed similar characteristics as mouse PGCs regarding morphology and ultrastructure, SSEA-1 expression and AP activity [Moens et al., 1997]. Injection of rabbit PGCs after short-term culture with human LIF and bFGF into embryos between the two-cell and expanded blastocyst stage resulted in low chimerism of the developing conceptuses and liveborn offspring. When nuclei from freshly isolated male and female rabbit PGCs were transferred into enucleated oocytes, blastocyst formation of male PGC-derived embryos was higher, but no implantation sites were detected following embryo transfer to recipients [Moens et al., 1996].

Rat

The in vitro propagation and chimera production with putative rat ES cells was first reported by Iannaccone et al. [1994]. Cells derived from blastocysts and cultured on rat embryonic fibroblasts with addition of LIF displayed several characteristics of pluripotent cells, such as SSEA-1 expression, AP activity, formation of embryoid bodies in vitro and of teratocarcinomas after injection into nude mice. However, the in vivo differentiation capacity of these cells (contribution to coat color chimeras) could not be confirmed by the same [Brenin et al., 1997] and other laboratories. Some rat ES-like cells cultivated for only 4 passages on homologous uterine epithelial cells were positive for AP activity and in some cases for ECMA-7 [Ouhibi et al., 1995]. In our own experiments [Vasilieva et al., unpubl. data] rat ES-like cell clones expressing AP activity, SSEA-1 and Oct-4 were established (fig. 1, B2, B3).

Syrian Hamster

Hamster putative ES cell lines were established from day 3 blastocysts and maintained undifferentiated on mouse primary embryonic fibroblast feeder layers for up to 25 passages [Doetschman et al., 1988]. In suspension, these cells formed embryoid bodies containing beating heart tissue, visceral yolk sac, endoderm cells, skeletal muscle cells and keratin swirls. However, contribution to germ line chimeras has still to be proven.

Mink

Morulae, immunosurgically isolated ICMs, or intact blastocysts have been used for the establishment of mink putative ES cell lines [Sukoyan et al., 1992, 1993]. The efficiency of suppressing differentiation and stimulating proliferation was low on primary mouse embryonic fibroblasts and established mink fibroblast cell lines [Sukoyan et al., 1992]. Morulae cultured on primary mink embryonic fibroblasts in the presence of LIF gave rise to ES-like cells capable of forming embryoid bodies in vitro and teratocarcinomas following injection into athymic mice, but these cells failed to contribute to chimeras after blastocyst injection [Sukoyan et al., 1993]. Another approach using mink blastocysts obtained after prolactin-induced termination of diapause resulted in AP-positive ES-like cells after 8 months in vitro [Polejaeva et al., 1997].

Pig

Since the pig is an excellent model for studying various human diseases, there are numerous reports concerning the culture and characterization of ICM- and PGC-derived cell lines. Embryos were recovered at the desired age ranging from day 6 to 9 [Evans et al., 1990; Notarianni et al., 1990; Piedrahita et al., 1990a,b; Wheeler, 1994; Gerfen and Wheeler, 1995; Ropeter-Scharfenstein et al., 1996] and also at more advanced stages [Strojek et al., 1990; Hochereau-de Reviers and Perreau, 1993; Anderson et al., 1994; Wianny et al., 1997]. Advanced day 10 and 11 embryos appear to provide the best initial outgrowths, but tend to differentiate in vitro, while ICMs from earlier stages yield less ES cell-like colonies with a reduced tendency to differentiate during prolonged culture. However, only freshly isolated ICM cells gave rise to viable porcine chimeras after blastocyst injection [Anderson et al., 1994].

The difficulties in establishing porcine ES cells may be due to differences in embryo development, notably in the prolonged period between hatching and implantation characterized by a marked elongation of the blastocyst [Geisert et al., 1982]. During this period the ICM differentiates into the epithelial embryonic disc [Gerfen and Wheeler, 1995] which does not substantially proliferate until gastrulation, but delaminates an inner layer of primitive endoderm just after hatching [Notarianni et al., 1991] and forms vimentin expressing mesoderm-like cells [Prelle et al., 1995b]. The rapidly outgrowing trophoblast as well as the early onset of ICM differentiation require a separation of the ICM from differentiation-inducing effects prior to in vitro culture using immunosurgery [Solter and Knowles, 1975] or calcium ionophore [Prelle et al., 1993].

Much effort has been spent to optimize culture conditions supporting proliferation of undifferentiated porcine embryonic cells [Vasiliev and Vasilieva, 1995]. While homologous porcine uterine fibroblasts supported embryonic attachment and colony formation for a short time [Strojek et al., 1990], undifferentiated ICM-derived cells could be maintained on STO feeder cells for more than 1 year. Injection of the ICM-derived cells into host blastocysts resulted in the development of two chimeric porcine fetuses which were retrieved at day 30 of pregnancy [Notarianni et al., 1997]. BRL cell-conditioned medium was suitable for propagating limited undifferentiated proliferation of porcine ES-like cells [Wheeler, 1994]; however, recombinant human LIF [Moore and Piedrahita, 1997] which is highly homologous to the corresponding porcine cytokine [Wianny et al., 1997], and other heterologous cytokines were ineffective in maintaining cell lines for extended periods of time [Prelle et al., 1994, 1995a] without loosing the capability of germ line transmission [Moore and Piedrahita, 1996].

Most of the porcine embryo-derived cell lines were capable of differentiating in vitro to some extent [Evans et al., 1990; Notarianni et al., 1990] and formed teratocarcinomas after injection into nude mice [Hochereau-de Reviers and Perreau, 1993]. Porcine ES-like cells showed AP activity [Talbot et al., 1993], expressed SSEA-1 [Wianny et al., 1997], but – in contrast to murine ES cells – displayed an epithelial phenotype characterized by cytokeratin expression [Piedrahita et al., 1990b]. Recently, putative porcine ES cells displaying a morphology similar to mouse ES cells were maintained in culture for up to 44 passages and, after injection into blastocysts, contribute to chimeric offspring as assessed by the pattern of coat color and by using a genetic marker, while germ line transmission still has to be proven [Wheeler, 1994]. In vitro derived embryoid bodies originating from these cells differentiated into fibroblasts, epithelium, neurons and muscle cells [Gerfen and Wheeler, 1995].

Porcine PGCs were isolated from fetal genital ridges between days 24 and 26 post conception [Takagi et al., 1997; Piedrahita et al., 1998], but also earlier during their migration across the dorsal mesentery. In this latter stage PGCs are characterized by a higher proliferation rate and by a reduced tendency to differentiate [Shim et al., 1997]. Porcine PGCs which express SSEA-1 in vivo [Takagi et al., 1997] require STO feeder cells [Shim and Anderson, 1998], LIF, porcine SCF, but – in contrast to mouse – no bFGF for initial culture [Durcova-Hills et al., 1998]. For prolonged undifferentiated proliferation they are more responsive to cell membrane-bound [Takagi et al., 1997; Durcova-Hills et al., 1998] or homologous cytokines [Piedrahita et al., 1997].

Recently reported porcine EG cells morphologically resemble murine ES cells and consistently express AP activity [Shim et al., 1997]. These cells maintained a normal diploid karyotype, were capable of in vitro differentiation, and contributed to tissues of a liveborn chimeric piglet which, however, died after a few days. Blastocyst injection of transfected PGCs carrying the green fluorescent protein gene resulted in the birth of a chimeric piglet with contribution of transgenic cells [Piedrahita et al., 1998]. Others established several porcine EG cell lines expressing SSEA-1 and AP activity which contributed to different tissues of healthy chimeric piglets born after blastocyst injection of the transgenic EG cells [Müller et al., 1999] (fig. 1C).

Cattle

In cattle, the establishment of ES cell lines is more feasible than in the pig because large numbers of embryos can be produced in vitro; however, in vivo-derived blastocysts, especially early-hatching blastocysts, are a superior source of pluripotent cells [Talbot et al., 1995].

In cattle as in most other species the time when nuclei of embryonic cells become committed to a specific cell fate and lose their ability to direct development of different cell types limits the success of ES cell establishment [Keefer et al., 1994]. Undifferentiated cells have been isolated at any stage from 16-cell to hatched blastocyst [Strelchenko, 1996], from the latter stage mainly by immunosurgery [Sims and First, 1993; Prelle et al., 1996b]. The critical issue of getting a tight contact between feeder and embryonic cells was solved by placing the disaggregated cells under the monolayer [Stice et al., 1996].

A number of different feeder cell types have been used, including homologous granulosa cells [Strojek-Baunack et al., 1991] or fetal fibroblasts [First et al., 1994], murine fetal fibroblasts [Saito et al., 1992; Prelle et al., 1996a,

1997] and STO cells [Talbot et al., 1995]; however, none of these feeder cell types was suitable for stimulating proliferation of embryonic cells without differentiation for an extended period of time. Supplementation of the culture medium with the heterologous human LIF did not show an effect on bovine ICM cells [Saito et al., 1992], even though bovine LIF is more similar to human than murine LIF [First et al., 1994]. The embryonic cells differentiated into a wide variety of cell types including trophectoderm which – in contrast to other species – showed AP activity [Talbot et al., 1995] while undifferentiated bovine ES cells were AP-negative [Cibelli et al., 1998a].

Sims and First [1993] maintained dissociated ICM cells in microdrop suspension culture preventing cell-to-cell contact which is required for differentiation [First et al., 1994]. Cells cultured in this system for less than 28 days resulted in the birth of live calves when used as donors in nuclear transfer experiments. Stice et al. [1996] reported several ES-like cell lines which were evaluated by morphological criteria and by their ability to form embryoid bodies after more than 50 passages. Nuclear transfer embryos produced from these cells initiated pregnancies, however, they were aborted before day 60 due to incomplete placental development. This abnormality has been interpreted as a consequence of erased imprinting [Strelchenko, 1996] and could be prevented by aggregation of ES-like cells with tetraploid embryos which contribute almost exclusively to the placental tissue [Stice et al., 1996].

Cibelli et al. [1998a] established putative bovine ES-like cells which proliferated for more than 12 months without differentiation and displayed ES cell characteristics (high nuclear:cytoplasmic ratio, defined colony margins). In one approach blastocyst outgrowth was cultured and embryo-derived cells were transfected by DNA microinjection of a β-galactosidase (β-Gal)-neomycin (β-Geo) expression vector. The same authors transfected bovine fetal fibroblasts with β-Geo by DNA electroporation and used these cells as nuclear donors to produce transgenic blastocyst stage nuclear transfer embryos to derive ES-like cells from their ICM. Morula injection of either the embryo- or nuclear transfer-derived transgenic ES-like cells resulted in the birth of nine chimaeric calves carrying the β-Gal transgene in at least one of the tissues investigated.

In cattle, several attempts to establish pluripotent cell lines from PGCs took advantage of the cross-species similarity observed in the patterns of migration and proliferation [Leichthammer et al., 1990]. Genital ridges were obtained from fetuses between days 29 and 70 post conception [Cherny et al., 1994; Lavoir et al., 1994; Strelchenko, 1996; Lavoir et al., 1997] and around day 175 of gestation [Delhaise et al., 1995]. The PGCs were isolated either by enzymatic digestion or by mechanic disruption of the recovered tissues and identified by size and other morphological criteria such as large nucleus, intracytoplasmic vesicles, and occasional blebbing [Leichthammer et al., 1990]. The cells were cultured either on primary embryonic fibroblasts or on feeder layers expressing recombinant human LIF [Cherny et al., 1994], and the medium was supplemented with a 'cocktail' of human SCF, recombinant LIF and bovine bFGF [Strelchenko, 1996]. Pluripotency was evaluated by formation of embryoid bodies in vitro [Cherny et al., 1994] and by injection into blastocysts, demonstrating that FITC-labelled PGCs integrated into the ICM of a chimeric embryo [Cherny et al., 1994]. After nuclear transfer of freshly isolated bovine PGCs, developing embryos initiated pregnancies until day 40 [Delhaise et al., 1995] and 43 [Lavoir et al., 1997], respectively. Recently, the birth of one healthy calf after nuclear transfer of cultured PGCs and recloning of the developing morulae was announced [Strelchenko et al., 1998], however, another calf derived directly from freshly isolated PGCs died soon after delivery [Zakhartchenko et al., 1999a].

Sheep

For the establishment of ovine ES cells, embryos were collected between days 7 and 13 post conception [Handyside et al., 1987; Notarianni et al., 1991; Talbot et al., 1993; Galli et al., 1994; Karasiewicz et al., 1996; Meinecke-Tillmann and Meinecke, 1996]; however, the proportion of attachment to feeder cells increased when using advanced embryonic stages [Meinecke-Tillmann and Meinecke, 1996]. Isolated ICMs or embryonic discs [Notarianni et al., 1991] showed a more frequent and faster occurring attachment than intact embryos [Piedrahita et al., 1990b]. Ovine ES-like cells do not form multilayered colonies, but grow as epithelioid monolayers [Notarianni et al., 1991]. Pluripotency of the cells could be maintained only for a few passages, as shown by their ability to differentiate in vitro [Piedrahita et al., 1990b; Meinecke-Tillmann and Meinecke, 1996] and by AP activity [Talbot et al., 1993; Galli et al., 1994]. Ovine embryonic cells proliferated and differentiated relatively independent of the feeder cell type [Piedrahita et al., 1990a]. STO and BRL cells [Handyside et al., 1987], a range of species-homologous and heterologous bovine fetal feeder cell types [Meinecke-Tillmann and Meinecke, 1996], and supplementation of the culture medium with LIF were not sufficient to prevent spontaneous differentiation into endoderm-like [Handyside et al., 1987], mesenchymal [Karasiewicz et al., 1996] or epithelial cells [Campbell et al., 1996a]. Nevertheless the latter cells led to birth of live lambs after use in nuclear transfer experiments [Campbell et al., 1996a; Wells et al., 1997]. Attempts to prove the pluripotency of ovine embryo-derived cell lines by in vivo differentiation after injection into nude mice [Galli et al., 1994] or by chimera formation and germ line transmission after blastocyst injection [Handyside et al., 1987] failed.

Nonhuman Primates

Because of the differences between human and mouse development, anatomy, and physiology, primate models would increase the understanding of the pathogenesis of specific human diseases, such as Alzheimer's and Parkinson's disease. The cultivation of immunosurgically isolated ICMs of rhesus monkey blastocysts on a mouse embryonic fibroblast feeder layer in LIF-supplemented medium gave rise to ES-like cells which could be maintained in an undifferentiated state for more than 1 year [Thomson et al., 1995]. These cells resembled human embryonic carcinoma cells with respect to AP activity and characteristic surface markers (SSEA-3 and SSEA-4). In the absence of feeder cells, these cells differentiated into multiple cell types, and injection into immunodeficient mice resulted in teratocarcinoma formation. Eight pluripotent cell lines derived from day 8 embryos of the common marmoset [Thomson et al., 1996] were maintained in culture for more than 12 months and displayed morphological features similar to those of embryo-derived cells of rhesus monkey. Each cell line had a normal karyotype, was able to differentiate in vitro into various cell types, and formed embryoid bodies.

ES Cells and Genetic Modification

Originally, ES cells have been used for the study of early developmental processes, e.g. the clonal analysis of cell lineage in mammals, and the detection of stage- and cell-specific gene expression [Anderson, 1992]. ES cell-mediated transgenesis has some distinct advantages over other transgenic methods such as integration of (retro-)viral vectors into an early embryo or microinjection of DNA into the pronucleus of a fertilized oocyte with the difficulties of random integration of multiple copies of the introduced DNA [reviewed in Stewart, 1991]. ES cell technology might significantly increase the efficiency of producing transgenic animals, as individual cell lines derived from a single transfected cell can be screened in vitro for integration and expression of the exogenous DNA construct before creating germ line chimeric animals [Wheeler, 1994]. Because of the rapid proliferation in vitro, ES cells provide an inexhaustible supply of cells capable of homologous recombination [Capecchi, 1989]. This technique allows the precise modification of existing genes, overcomes the problems of positional effects and insertional inactivation, and mediates the inactivation of specific endogenous genes [Osterrieder and Wolf, 1998]. The reinsertion of the modified ES cell genome within the germ line provides an almost abundant source of transgenic animals [Seamark, 1994], and a method to rapidly introduce new genetic information without conventional cross-breeding [Wheeler and Choi, 1997]. Gene transfer focuses on the improvement of productivity traits (milk and carcass composition, growth rate and feed utilization) but also of disease resistance and reproductive performance [Wheeler et al., 1995]. Other important applications include the production of pharmaceutical proteins in specific organs or body fluids of transgenic animals (gene farming) [Brem and Müller, 1994], and the genetic modification of pigs to produce cells, tissues or organs for xenotransplantation [reviewed in Platt, 1998]. In addition, genetically engineered farm animals may become important animal models. The establishment of animal homologues for human diseases has proven to be an invaluable tool for studies of human genetic and infectious disorders and cancer. In particular, pigs offer some distinct advantages compared to the commonly used rodent models because they are immunologically and physiologically more similar to humans and have a more diverse genetic background [Brem and Müller, 1994; Wheeler, 1994]. Recent progress in nuclear transfer using transfected cells offers an alternative, apparently more efficient strategy for producing transgenic animals than DNA microinjection, especially in species where ES cells are not established but nuclear transfer is successfully performed (fig. 2).

Potential Benefits of Totipotent and Somatic Cell Cloning

Cloning is an asexual form of reproduction, resulting in genetically identical organisms [Anderson and Seidel, 1998]. Embryo cloning by nuclear transfer involves the transfer of genetic material from a donor cell (karyoplast) to the cytoplasm of an oocyte or zygote from which the genetic material has been removed (cytoplast). Since the

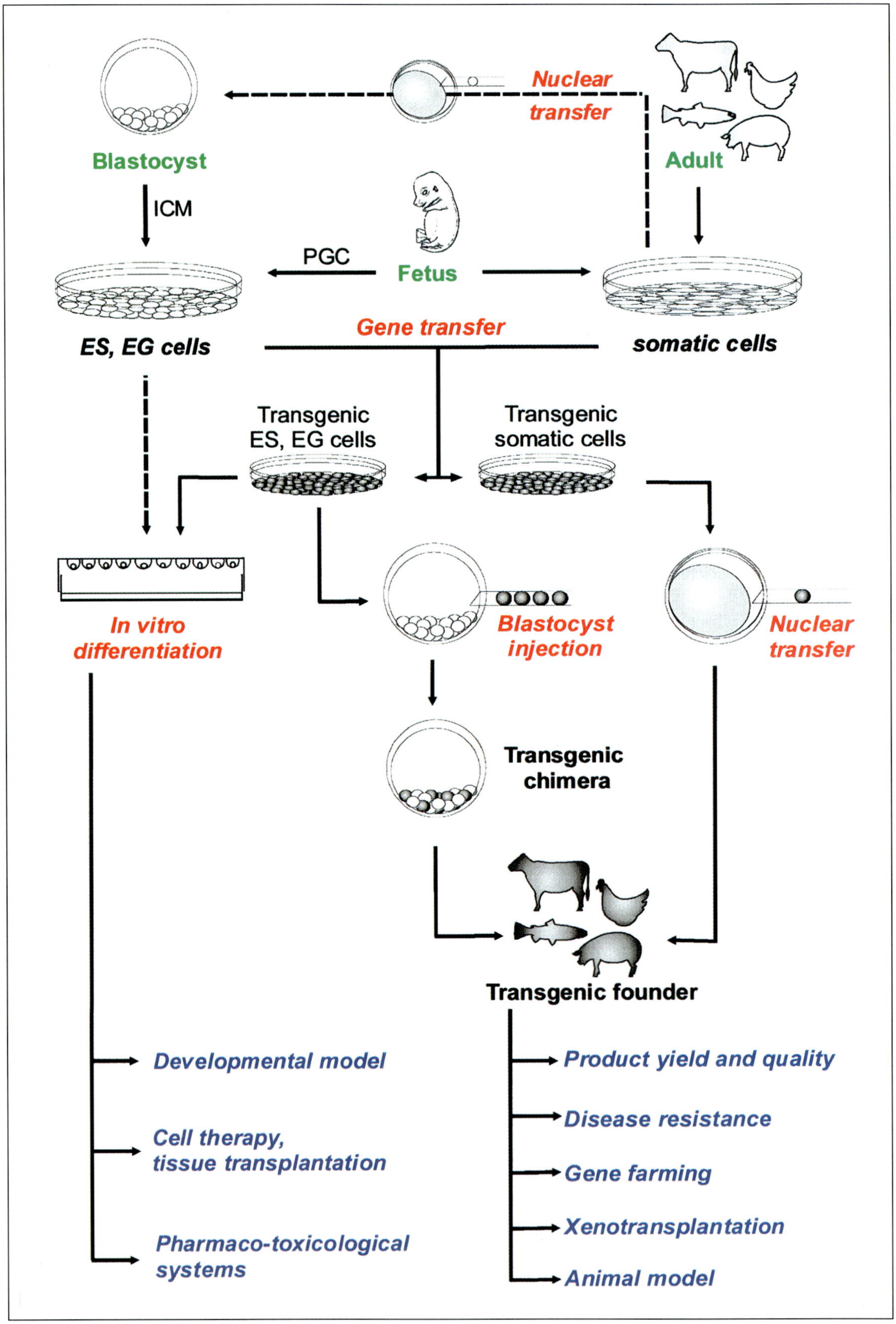
Nuclear
transfer
Blastocyst
Adult
ICM
PGC
Fetus
ES, EG cells
Gene transfer
somatic cells
Transgenic
ES, EG cells
Transgenic
somatic cells
In vitro
differentiation
Blastocyst
injection
Nuclear
transfer
Transgenic
chimera
Transgenic founder
Developmental model
Cell therapy,
tissue transplantation
Pharmaco-toxicological
systems
Product yield and quality
Disease resistance
Gene farming
Xenotransplantation
Animal model

number of cells in preimplantation embryos is limited, their use as donor nuclei prohibits the generation of larger clones and also genetic modification. Recent progress in cloning animals from cultured cells of embryonic, fetal, or adult origin offers the possibility of unlimited multiplication of elite embryos or animals, and of precise genetic modification [reviewed in Wolf et al., 1998] (fig. 2). With the conventional use of ES cells via blastocyst injection or morula aggregation chimeric animals are generated and subsequent production of homozygous transgenic founder animals by conventional cross-breeding is necessary [reviewed in Stice et al., 1998]. In contrast, after the utilization of transfected presumptive ES or somatic cells as donors in nuclear transfer experiments, all offspring of the first generation will consist entirely of donor cell nuclei carrying the transgene. Also the sex of the nuclear donor cells can be determined prior to use and the problem of mosaicism in the transgenic founders resulting in failure of transgene transmission in the gametes is eliminated [Anderson and Seidel, 1998]. Production of transgenic livestock by nuclear transfer of transfected cultured cells requires fewer embryos than direct DNA microinjection into zygotes [Schnieke et al., 1997]. This implies important advantages saving time and expense particularly in large domestic animals with a long generation time and a limited number of offspring [First and Thomson, 1998].

Additionally, the availability of cultured cells for nuclear transfer allows further investigation of the cell cycle and its role in the development of cloned embryos [Campbell et al., 1996b]. Some investigators argue that forcing the donor cell into a quiescent G0 stage enhances nuclear reprogramming by switching off genes in differentiated cells [Campbell et al., 1996b]. Reprogramming includes reduction or cessation of transcription, changes in the nuclear structure (nuclear lamina), chromatin structure, nuclear morphology and stage-specific protein synthesis [reviewed in Wolf et al., 1998]. Reprogramming is also affected by the timing of embryonic genome activation [reviewed in Stice et al., 1998], which occurs at a later stage in sheep and cattle (8-cell stage), but earlier in species less amenable to nuclear transfer (mouse: 2-cell stage; pig: 4-cell stage), leaving a shorter time period for the transferred nucleus to be reprogrammed [Solter, 1998].

Fig. 2. Pluripotent and somatic cells as alternatives for gene and nuclear transfer in vertebrates: derivation, genetic modification and utilization via in vitro differentiation, blastocyst injection and nuclear transfer for in vitro studies, cell therapy and production of transgenic animals for different purposes.

However, Wakayama et al. [1998] succeeded in full-term development of mice by simple nuclear transfer of cumulus cells. This report also underlines a series of findings initiated by Campbell et al. [1996a] suggesting that nuclei from differentiated cells introduced into enucleated oocytes are capable of supporting full development. Originally, differentiated cells have been thought to be inappropriate as nuclear donors because of apparently irreversible changes in their genome during differentiation (e.g. epigenetic methylation of cytosines). This dogma has been overcome by successful nuclear transfer using fetal cells in cattle and sheep [Schnieke et al., 1997; Wells et al., 1997; Cibelli et al., 1998b; Zakhartchenko et al., 1999b]. In the first successful cloning experiment of goats donor karyoplasts were obtained from a transgenic female fetus and one of three identical offspring produced human antithrombin III in its milk [Baguisi et al., 1999]. Fully differentiated somatic cells from adult animals used as nuclear donors, e.g. oviductal cells [Kato et al., 1998] and mural granulosa cells [Wells et al., 1999], resulted in healthy calves. However, success might be limited due to decreased and variable telomere length in the latter [Anderson and Seidel, 1998]. Mitotically inactive adult cells, such as neurons, might be poor candidates, while continuously dividing cells, such as epidermal cells, should be preferred [Wilmut et al., 1997; Wakayama et al., 1998]. Also a random loss of correct imprinting and the inactive X chromosome in differentiated female cells may be responsible for low full-term development of embryos derived from adult cell nuclei [Solter, 1998]. Not only a high rate of perinatal mortality but long-lasting deleterious effects might be associated with adult somatic cloning. The random choice of cultured donor cells bearing mutations or the lack of fine-tuned regulation during nucleocytoplasmic interaction are considered as epigenetic events affecting the correct reprogramming of gene activity [Renard et al., 1999].

These difficulties together with a finite life span of differentiated cells in culture also limit the types of transgenic modifications that can be made. Homologous recombination requires the selection of transfected cells and the propagation of these cells to identify the correctly targeted ones. This common strategy would be limited by a low number of cell divisions [Cibelli et al., 1998b]. Therefore, the establishment of pluripotent cells still remains a matter of scientific interest, at least as a unique source for studies of early embryonic development in respect to cell commitment and gene regulation.

Potential Applications of ES Cells and Cell Cloning to Human Biology and Medicine

Although the recent success in establishing human ES cells raised numerous ethical concerns, acceptable applications of ES cell technology in the human can be envisaged and will be based on manipulation and analysis of human ES cells in vitro. Bongso et al. [1994] reported the isolation of human ICM cells at day 8 postinsemination and culture on ampullary epithelial feeder layers for 2 passages using human LIF. ES-like cells were identified by typical morphology and AP activity. Recently, five human ES cell lines were established from immunosurgically isolated ICMs, and at least one was maintained for more than 8 months showing similar morphology to rhesus monkey ES cells [Thomson et al., 1998]. ES-like character was demonstrated by surface marker (SSEA-3, SSEA-4) expression, AP activity, teratoma/teratocarcinoma formation in nude mice, and high telomerase activity (which is involved in maintenance of telomere length and correlates with immortality in human cell lines). Additionally, human primordial germ cells were derived from 5- to 9-week-old fetuses and cultured on STO cells with medium containing human LIF and bFGF for up to 21 days. The growing colonies were AP-positive, expressed characteristic immunological markers (SSEA-1, SSEA-3, SSEA-4), and differentiated in vitro into derivatives of all three germ layers [Shamblott et al., 1998]. In vitro differentiation of human ES cells directed along chosen pathways would allow investigation of human developmental events including regulatory signals, cell commitment and morphogenesis, and identification of target genes for new drugs and teratogenic or toxic compounds that cannot be analyzed in vivo due to ethical constraints [Rathjen et al., 1998; Thomson et al., 1998]. ES cell technology together with cloning can be useful in humans for cell and tissue therapy [Gearhart, 1998; Trounson and Pera, 1998] (fig. 2). People could provide their own cells and, by using them to replace the nuclei of their own or donor eggs, obtain stem cells in culture and then induce differentiation to provide individually tailored cell and tissue replacements without rejection problems offering lifelong treatment [Gage, 1998]. The major obstacles for gene therapy, (1) the lack of long-lived cell types easily isolated, maintained and manipulated in vitro, and (2) the possibilities of only crude modification of the human genome, could be solved by permanent human ES cell lines suitable for precise genetic modification [Rathjen et al., 1998]. But before future application of the human ES cell technology, international rules should be worked out with respect to ethical implications and the commercial use of human ES cells [see Anderson and Seidel, 1998].

Acknowledgements

We thank Dr. Sigrid Müller for providing the photographs of porcine PGCs, for many helpful comments and extended thoughtful discussions. Work was partially supported by DFG grants WO 685/2-1 and WO 685/3-1 to E.W., and by Boehringer Ingelheim and by the Fonds der Chemischen Industrie supporting A.M.W.

References

Anderson, G.B. (1992) Isolation and use of embryonic stem cells from livestock species. Anim Biotechnol *3*: 165–175.

Anderson, G.B., S.J. Choi, R.H. BonDurant (1994) Survival of porcine inner cell masses in culture and after injection into blastocysts. Theriogenology *42*: 204–212.

Anderson, G.B., G.E. Seidel (1998) Cloning for profit. Science *280*: 1400–1401.

Baguisi, A., E. Behboodi, D.T. Melican, J.S. Pollock, M.M. Destrempes, C. Cammuso, J.L. Williams, S.D. Nims, C.A. Porter, P. Midura, M.J. Palacios, S.L. Ayres, R.S. Denniston, M.L. Hayes, C.A. Ziomek, H.M. Meade, R.A. Godke, W.G. Gavin, E.W. Overström, Y. Echelard (1999) Production of goats by somatic cell nuclear transfer. Nat Biotechnol *17*: 456–461.

Beddington, R.S.P., E.J. Robertson (1989) An assessment of the developmental potential of embryonic stem cells in the mid-gestation embryo. Development *105*: 733–737.

Bongso, A., C-Y. Fong, S-C. Ng, S. Ratnam (1994) Isolation and culture of inner cell mass cells from human blastocysts. Hum Reprod *11*: 2110–2117.

Bradley, A., M. Evans, M.H. Kaufman, E. Robertson (1984) Formation of germ line chimeras from embryo-derived teratocarcinoma cell lines. Nature *309*: 255–256.

Brem, G., M. Müller (1994) Large transgenic animals; in Maclean, N. (ed): Animals with Novel Genes. Cambridge, Cambridge University Press, pp 179–244.

Brenin, D., J. Look, N. Hübner, G. Levan, P. Iannaccone (1997) Rat embryonic stem cells: A progress report. Transplant Proc *29*: 1761–1765.

Brook, F.A., R.L. Gardner (1997) The origin and efficient derivation of embryonic stem cells in the mouse. Proc Natl Acad Sci USA *94*: 5709–5712.

Campbell, K.H.S., J. McWhir, W.A. Ritchie, I. Wilmut (1996a) Sheep cloned by nuclear transfer from a cultured cell line. Nature *380*: 64–66.

Campbell, K.H.S., W.A. Ritchie, J. McWhir, I. Wilmut (1996a) Cloning farm animals by nuclear transfer: From cell cycle to cells. Emb Trans News *14*: 12–17.

Campbell, K.H.S., I. Wilmut (1997) Totipotency or multipotentiality of cultured cells: Applications and progress. Theriogenology *47*: 63–72.

Capecchi, M.R. (1989) Altering the genome by homologous recombination. Science *244*: 1288–1292.

Carsience, R.S., M.E. Clark, A.M. Verrinder Gibbins, R.J. Etches (1993) Germ line chimeric chickens from dispersed donor blastodermal cells and compromised recipient embryos. Development *117*: 669–675.

Chang, I.-K., D.L. Jeong, Y.H. Hong, T.S. Park, Y.K. Moon, T. Ohno, J.Y. Han (1997) Production of germ line chimeric chickens by transfer of cultured primordial germ cells. Cell Biol Int *21*: 495–499.

Cherny, R.A., T.M. Stokes, J. Merei, L. Lom, M.R. Brandon, R.L. Williams (1994) Strategies for the isolation and characterization of bovine embryonic stem cells. Reprod Fertil Dev *6*: 569–575.

Cibelli, J.B., S.L. Stice, P.G. Golueke, J.J. Kane, J. Jerry, E.S. C. Blackwell, F.A. Ponce de Leon, J.M. Robl (1998a) Transgenic bovine chimeric offspring produced from somatic cell-derived stem-like cells. Nat Biotechnol *16*: 642–646.

Cibelli, J.B., S.L. Stice, P.G. Golueke, J.J. Kane, J. Jerry, E.S.C. Blackwell, F.A. Ponce de Leon, J.M. Robl (1998b) Cloned transgenic calves produced from nonquiescent fetal fibroblasts. Science *280*: 1256–1258.

Collodi, P., Y. Kamei, A. Sharps, D. Weber, D. Barnes (1992) Fish embryo cell cultures for derivation of stem cells and transgenic chimeras. Mol Mar Biol Biotechnol *1*: 257–265.

Conover, J.C., N.Y. Ip, W.T. Poueymirou, B. Bates, M.P. Goldfarb, T.M. DeChiara, G.D. Yancopoulos (1993) Ciliary neutrophic factor maintains the pluripotentiality of embryonic stem cells. Development *119*: 559–565.

Delhaise, F., V. Bralion, N. Schuurbiers, F. Dessy (1996) Establishment of an embryonic stem cell line from 8-cell stage mouse embryos. Eur J Morphol *34*: 237–243.

Delhaise, F., F.J. Ectors, R. De Roover, F. Ectors, F. Dessy (1995) Nuclear transplantation using bovine primordial germ cells from male fetuses. Reprod Fertil Dev *7*: 1217–1219.

Doetschman, T.C., H.R. Eistetter, M. Katz, W. Schmidt, R. Kemler (1985) The in vitro development of blastocyst-derived embryonic stem cell lines: Formation of visceral yolk sac, blood islands and myocardium. J Embryol Exp Morphol *87*: 27–45.

Doetschman, T.C., P. Williams, N. Maeda (1988) Establishment of hamster blastocyst-derived embryonic stem (ES) cells. Dev Biol *127*: 224–227.

Du, F., J.R. Giles, R.H. Foote, K.H. Graves, X. Yang, R.W. Moreadith (1995) Nuclear transfer of putative rabbit embryonic stem cells leads to normal blastocyst development. J Reprod Fertil *104*: 219–223.

Durcova-Hills, G., K. Prelle, S. Müller, M. Stojkovic, J. Motlik, E. Wolf, G. Brem (1998) Primary culture of porcine PGCs requires LIF and porcine membrane-bound stem cell factor. Zygote *6*: 271–275.

Eddy, E.M., J.M. Clark, D. Gong, B.A. Fenderson (1981) Origin and migration of primordial germ cells in mammals. Gamete Res *4*: 333–362.

Eistetter, H.R. (1989) Pluripotent embryonal stem cell lines can be established from disaggregated mouse morulae. Dev Growth Differ *31*: 275–282.

Etches, R.J., M.E. Clark, A. Toner, G. Liu, A.M. Verrinder Gibbins (1996) Contribution to somatic and germ line lineage of chicken blastodermal cells maintained in culture. Mol Reprod Dev *45*: 291–298.

Evans, M.J., M.B.L. Carlton, A.P. Russ (1997) Gene trapping and functional genomics. Trends Genet *13*: 370–374.

Evans, M.J., M.H. Kaufman (1981) Establishment in culture of pluripotential cells from mouse embryos. Nature *292*: 154–156.

Evans, M.J., E. Notarianni, S. Laurie, R.M. Moor (1990) Derivation and preliminary characterization of pluripotent cell lines from porcine and bovine blastocysts. Theriogenology *33*: 125–128.

First, N.L., M.M. Sims, S.P. Park, M.J. Kent-First (1994) Systems for production of calves from cultured bovine embryonic cells. Reprod Fertil Dev *6*: 553–562.

First, N.L., J. Thomson (1998) From cows stem therapies? Nat Biotechnol *16*: 620–621.

Gage, F.H. (1998) Cell therapy. Nature *392(suppl)*: 18–24.

Galli, C., G. Lazzari, J.E. Flechon, R.M. Moor (1994) Embryonic stem cells in farm animals. Zygote *2*: 385–389.

Gearhart, J. (1998) New potential for human embryonic stem cells. Science *282*: 1061–1062.

Geisert, R.D., J.W. Brookbank, R.M. Roberts, F.W. Bazer (1982) Establishment of pregnancy in the pig. II. Cellular remodelling of the porcine blastocyst during elongation on day 12 of pregnancy. Biol Reprod *27*: 941–955.

Gerfen, R.W., M.B. Wheeler (1995) Isolation of embryonic cell lines from porcine blastocysts. Anim Biotechnol *6*: 1–14.

Giles, J.R., X. Yang, W. Mark, R.H. Foote (1993) Pluripotency of cultured rabbit inner cell mass cells detected by isozyme analysis and eye pigmentation of fetuses following injection into blastocysts or morulae. Mol Reprod Dev *36*: 130–138.

Graves, K.H., R.W. Moreadith (1993) Derivation and characterisation of putative pluripotential embryonic stem cells from preimplantation rabbit embryos. Mol Reprod Dev *36*: 424–433.

Handyside, A., M.L. Hooper, M.H. Kaufman, I. Wilmut (1987) Towards the isolation of embryonal stem cell lines. Rouxs Arch Dev Biol *196*: 185–190.

Hochereau-de Reviers, T.M., C. Perreau (1993) in vitro culture of embryonic disc cells from porcine blastocysts. Reprod Nutr Dev *33*: 475–483.

Hong, Y., M. Schartl (1996) Establishment and growth responses of early medakafish *(Oryzias latipes)* embryonic cells in feeder layer-free cultures. Mol Mar Biol Biotechnol *5*: 93–104.

Hong, Y., C. Winkler, M. Schartl (1996) Pluripotency and differentiation of embryonic stem cell lines from the medakafish *(Oryzias latipes)*. Mech Dev *60*: 33–44.

Hong, Y., C. Winkler, M. Schartl (1998) Production of medakafish chimeras from a stable embryonic stem cell line. Proc Natl Acad Sci USA *95*: 3679–3684.

Iannaccone, P.M., G.U. Taborn, R.L. Garton, M.D. Caplice, R.D. Brenin (1994) Pluripotent embryonic stem cells from the rat are capable of producing chimeras. Dev Biol *163*: 288–292.

Karasiewicz, J., E. Szablisty, A. Guszkiewicz, M. Kossakowski, G. Stefanski, M. Reed, J. Modlinski (1996) Development of isolated sheep inner cell masses/embryonic discs in vitro. Rouxs Arch Dev Biol *205*: 437–442.

Kato, Y., T. Tani, Y. Sotomaru, K. Kurokawa, J. Kato, H. Doguchi, H. Yasue, Y. Tsunoda (1998) Eight calves cloned from somatic cells of a single adult. Science *282*: 2095–2098.

Keefer, C.L., S.L. Stice, D.L. Matthews (1994) Bovine inner cell mass cells as donor nuclei in the production of nuclear transfer embryos and calves. Biol Reprod *50*: 935–939.

Kuwana, T., K. Hashimoto, A. Nakanishi, Y. Yasuda, A. Tajima, M. Naito (1996) Long-term culture of avian embryonic cells in vitro. Int J Dev Biol *40*: 1061–1064.

Labosky, P.A., D.P. Barlow, B.M.L. Hogan (1994) Mouse embryonic germ (EG) cell lines: Transmission through the germ line and differences in the methylation imprint of insulin-like growth factor 2 receptor *(Igf2r)* gene compared with embryonic stem (ES) cell lines. Development *120*: 3197–3204.

Lavoir, M.-C., P.K. Basrur, K.J. Betteridge (1994) Isolation and identification of germ cells from fetal bovine ovaries. Mol Reprod Dev *37*: 413–424.

Lavoir, M.-C., N. Rumph, A. Moens, W.A. King, Y. Plante, W.H. Johnson, J. Ding, K.J. Betteridge (1997) Development of bovine nuclear transfer embryos made with oogonia. Biol Reprod *56*: 194–199.

Leichthammer, F., E. Baunack, G. Brem (1990) Behaviour of living primordial germ cells of livestock in vitro. Theriogenology *33*: 1221–1230.

Mann, J.R., I. Gadi, M.L. Harbison, S.J. Abbondanzo, C.L. Stewart (1990) Androgenetic mouse embryonic stem cells are pluripotent and cause skeletal defects in chimeras: Implications for genetic imprinting. Cell *62*: 251–260.

Martin, G. (1981) Isolation of a pluripotent cell line from early mouse embryos cultured in medium conditioned by teratocarcinoma stem cells. Proc Natl Acad Sci USA *78*: 7634–7638.

Martin, G.R., M.J. Evans (1975) Differentiation of clonal lines of teratocarcinoma cells: Formation of embryoid bodies in vitro. Proc Natl Acad Sci USA *72*: 1441–1445.

Matsui, Y., K. Zsebo, B.L.M. Hogan (1992) Derivation of pluripotential embryonic stem cells from murine primordial germ cells in culture. Cell *70*: 841–847.

McWhir, J., A.E. Schnieke, R. Ansell, H. Wallace, A. Colman, A.R. Scott, A.J. Kind (1996) Selective ablation of differentiated cells permits isolation of embryonic stem cell lines from murine embryos with a non-permissive genetic background. Nat Genet *14*: 223–226.

Meinecke-Tillmann, S., B. Meinecke (1996) Isolation of ES-like cell lines from ovine and caprine pre-implantation embryos. J Anim Breed Genet *113*: 413–426.
Modlinski, J.A., M.A. Reed, T.E. Wagner, J. Karasiewicz (1996) Embryonic stem cells: Developmental capabilities and their possible use in mammalian embryo cloning. Anim Reprod Sci *42*: 437–446.
Moens, A., S. Chastant, P. Chesne, J.-E. Flechon, K.J. Betteridge, J.-P. Renard (1996) Differential ability of male and female rabbit fetal germ cell nuclei to be reprogrammed by nuclear transfer. Differentiation *60*: 339–345.
Moens, A., B. Flechon, J. Degrouard, X. Vignon, J. Ding, J.-E. Flechon, K.J. Betteridge, J.-P. Renard (1997) Ultrastructural and immunocytochemical analysis of diploid germ cells isolated from fetal rabbit gonads. Zygote *5*: 47–60.
Moore, K., J.A. Piedrahita (1996) Effects of heterologous hematopoietic cytokines on in vitro differentiation of cultured porcine inner cell masses. Mol Reprod Dev *45*: 139–144.
Moore, K., J.A. Piedrahita (1997) The effects of human leukemia inhibitory factor (HLIF) and culture medium on in vitro differentiation of cultured porcine inner cell mass (PICM). In Vitro Cell Dev Biol Anim *33*: 62–71.
Müller, S., K. Prelle, N. Rieger, C. Lassnig, U. Luksch, H. Petznek, B. Aigner, M. Baetscher, E. Wolf, M. Müller, G. Brem (1999) Chimeric pigs following blastocyst injection of transgenic porcine primordial germ cells. Mol Reprod Dev, in press.
Nagy, A., E. Gocza, E.M. Diaz, V.R. Prideaux, E. Ivanyi, M. Markkula, J. Rossant (1990) Embryonic stem cells alone are able to support fetal development in the mouse. Development *110*: 815–821.
Nichols, J., I. Chambers, A. Smith (1994) Derivation of germ line competent embryonic stem cells with combination of interleukin-6 and soluble interleukin-6 receptor. Exp Cell Res *215*: 237–239.
Nichols, J.F., E.P. Evans, A.G. Smith (1990) Establishment of germ line-competent embryonic stem (ES) cells using differentiation inhibiting activity. Development *110*: 1341–1348.
Notarianni, E., C. Galli, S. Laurie, R.M. Moor, M.J. Evans (1991) Derivation of pluripotent, embryonic cell lines from the pig and sheep. J Reprod Fertil Suppl *43*: 255–260.
Notarianni, E., S. Laurie, R.M. Moor, M.J. Evans (1990) Maintenance and differentiation in culture of pluripotential embryonic cell lines from pig blastocysts. J Reprod Fertil Suppl *41*: 51–56.
Notarianni, E., S. Laurie, A. Ng, K. Sathasivam (1997) Incorporation of cultured embryonic cells into transgenic and chimeric porcine fetuses. Int J Dev Biol *41*: 537–540.
Osterrieder, N., E. Wolf (1998) Lessons from gene knockouts. Rev Sci Tech *17*: 351–364.
Ouhibi, N., N.F. Sullivan, J. English, W.H. Colledge, M.J. Evans, N.J. Clark (1995) Initial culture behaviour of rat blastocysts on selected feeder cell lines. Mol Reprod Dev *40*: 311–324.
Pain, B., P. Chenevier, J. Samarut (1999) Chicken ES cells and transgenic strategies. Cells Tissues Organs *165*: 212–219.
Pain, B., M.E. Clark, M. Shen, H. Nakazawa, M. Sakura, J. Samarut, R.J. Etches (1996) Long-term in vitro culture and characterisation of avian embryonic stem cells with multiple morphogenic potentialities. Development *122*: 2339–2348.
Pease, S., P. Braghetta, D. Gearing, D. Grail, R.L. Williams (1990) Isolation of embryonic stem (ES) cell lines in media supplemented with recombinant leukemia factor (LIF). Dev Biol *141*: 344–352.
Pedersen, R.A. (1994) Studies of in vitro differentiation with embryonic stem cells. Reprod Fertil Dev *6*: 543–552.
Petitte, J.N., M.E. Clark, G. Liu, A.M. Verrinder Gibbins, R.J. Etches (1990) Production of somatic and germ line chimeras in the chicken by transfer of early blastodermal cells. Development *108*: 185–189.
Piedrahita, J.A., G.B. Anderson, R.H. BonDurant (1990a) Influence of feeder layer type on the efficiency of isolation of porcine embryo-derived cell lines. Theriogenology *34*: 865–877.
Piedrahita, J.A., G.B. Anderson, R.H. BonDurant (1990b) On the isolation of embryonic stem cells: A comparative behaviour of murine, porcine and ovine embryos. Theriogenology *34*: 879–901.
Piedrahita, J.A., K. Moore, C. Lee, B. Oetama, R. Weaks, J. Ramsoondar, J. Thomson, J. Vasquez (1997) Advances in the generation of transgenic pigs via embryo-derived and primordial germ cell-derived cells. J Reprod Fertil Suppl *52*: 245–254.
Piedrahita, J.A., K. Moore, B. Oetama, C.-K. Lee, N. Scales, J. Ramsoondar, F.W. Bazer, T. Ott (1998) Generation of transgenic porcine chimeras using primordial germ cell-derived colonies. Biol Reprod *58*: 1321–1329.
Platt, J.L. (1998) New directions for organ transplantation. Nature *392(suppl)*: 11–17.
Polejaeva, I.A., W.A. Reed, T.D. Bunch, L.C. Ellis, K.L. White (1997) Prolactin-induced termination of obligate diapause of mink *(Mustela vison)* blastocysts in vitro and subsequent establishment of embryonic stem-like cells. J Reprod Fert *109*: 229–230.
Prelle, K., H. Füllgrabe, W. Holtz (1993) Isolation of the inner cell mass (ICM) for the establishment of embryonic stem cells (abstract). J Reprod Fertil Abstr Ser *12*: 6.
Prelle, K., W. Holtz, M. Osborn (1995a) The intermediate filament protein vimentin as differentiation marker in preimplantation porcine embryos (abstract). Biol Reprod *52(suppl)*: 177.
Prelle, K., F. Sinowatz, E. Wolf (1997) Effects of different thiol compounds on the ultrastructure of inner cell mass cells of bovine embryos (abstract). Theriogenology 47: 244.
Prelle, K., M. Stojkovic, G. Brem, E. Wolf (1996a) Isolation of bovine inner cell masses for the establishment of pluripotent cell lines (abstract). Proc 13th Int Congr Anim Reprod, P21-6.
Prelle, K., M. Stojkovic, M. Brielmeier, E. Wolf (1996b) Beneficial effects of thiol compounds on hatching rate and ICM outgrowth of bovine IVM/IVF embryos (abstract). Biol Reprod *54(suppl)*: 172.
Prelle, K., A.M. Wobus, W. Holtz (1995b) Porcine inner cell masses grow undifferentiated in the presence of STO cells and bFGF (abstract). J Reprod Fertil Abstr Ser *15*: 72.
Prelle, K., A.M. Wobus, E. Wolf, N. Neubert, W. Holtz (1994) Effects of growth factors on the in vitro development of porcine inner cell masses isolated by calcium ionophore A23187 (abstract). J Reprod Fertil Abstr Ser *13*: 41.
Rathjen, P.D., J. Lake, L.M. Whyatt, M.D. Bettess, J. Rathjen (1998) Properties and uses of embryonic stem cells: Prospects for application to human biology and gene therapy. Reprod Fertil Dev *10*: 31–47.
Renard, J.-P., S. Chastant, P. Chesne, C. Richard, J. Marchal, N. Cordonnier, P. Chavatte, X. Vignon (1999) Lymphoid hypoplasia and somatic cloning. Lancet *353*: 1489–1491.
Resnick, J.L., L.S. Bixler, L. Cheng, P.L. Donovan (1992) Long-term proliferation of mouse primordial germ cells in culture. Nature *359*: 550–551.
Robertson, E.J. (1987) Embryo-derived stem cell lines; in Robertson EJ (ed): Teratocarcinomas and Embryonic Stem Cells: A Practical Approach. Oxford, IRL Press, pp 71–112.
Robertson, E.J. (1997) Derivation and maintenance of embryonic stem cell cultures. Methods Mol Biol *75*: 173–184.
Rohwedel, J., U. Sehlmeyer, J. Shan, A. Meister, A.M. Wobus (1996) Primordial germ cell-derived mouse embryonic germ (EG) cells in vitro resemble undifferentiated stem cells with respect to differentiation capacity and cell cycle distribution. Cell Biol Int *20*: 579–587.
Rohwedel, J., K. Guan, A.M. Wobus (1999) Induction of cellular differentiation by retinoic acid in vitro. Cells Tissues Organs *165*: 190–202.
Ropeter-Scharfenstein, M., N. Neubert, K. Prelle, W. Holtz (1996) Identification, isolation, and culture of pluripotent cells from the porcine inner cell mass. J Anim Breed Gen *113*: 427–436.
Rose, T.M., G. Bruce (1994) Oncostatin M is a member of a cytokine family that includes leukaemia inhibitory factor, granulocyte colony stimulating factor, and interleukin-6. Proc Natl Acad Sci USA *88*: 8641–8645.
Saito, S., N. Strelchenko, H. Niemann (1992) Bovine embryonic stem cell-like lines cultured over several passages. Rouxs Arch Dev Biol *201*: 134–141.
Schnieke, A.E., A.J. Kind, W.R. Ritchie, K. Mycock, A.R. Scott, M. Ritchie, I. Wilmut, A. Colman, K.S. Campbell (1997) Human factor IX transgenic sheep produced by transfer of nuclei from transfected fetal fibroblasts. Science *278*: 2130–2133.
Schöler, H.R., A.K. Hatzopoulos, R. Balling, N. Suzuki, P. Gruss (1989) A family of octamer specific proteins present during mouse embryogenesis: Evidence for germ line-specific expression of an Oct factor. EMBO J *8*: 2543–2550.

Schoonjans, L., G.M. Albricht, J.-L. Li, D. Collen, R.W. Moreadith (1996) Pluripotential rabbit embryonic stem (ES) cells are capable of forming overt coat colour chimeras following injection into blastocysts. Mol Reprod Dev *45*: 439–443.

Seamark, R.F. (1994) Progress and emerging problems in livestock transgenesis: A summary perspective. Reprod Fertil Dev *6*: 653–657.

Shamblott, M.J., J. Axelman, S. Wang, E.M. Bugg, J.W. Littlefield, P.J. Donovan, P.D. Blumenthal, G.R. Huggins, J.D. Gearhart (1998) Derivation of pluripotent stem cells from cultured human primordial germ cells. Proc Natl Acad Sci USA *95*: 13726–13731.

Shim, H., G.B. Anderson (1998) In vitro survival and proliferation of porcine primordial germ cells. Theriogenology *49*: 521–528.

Shim, H., A. Gutierrez-Adan, L.R. Chen, R.H. BonDurant, E. Behboodi, G.B. Anderson (1997) Isolation of pluripotent stem cells from cultured porcine primordial germ cells. Biol Reprod *57*: 1089–1095.

Sims, M., N.L. First (1993) Production of calves by transfer of nuclei from cultured inner cell mass cells. Proc Natl Acad Sci USA *90*: 6143–6147.

Smith, A.G., J.K. Heath, D.D. Donaldson, G.G. Wong, J. Moreau, M. Stahl, D. Rogere (1988) Inhibition of pluripotential embryonic stem cell differentiation by purified polypeptides. Nature *336*: 688–690.

Smith, A.G., M.L. Hooper (1987) Buffalo rat liver cells produce a diffusible activity which inhibits the differentiation of murine embryonal carcinoma and embryonic stem cells. Dev Biol *121*: 1–9.

Solter, D. (1998) Dolly is a clone – and no longer alone. Nature *394*: 315–316.

Solter, D., B.B. Knowles (1975) Immunosurgery of mouse blastocyst. Proc Natl Acad Sci USA *72*: 5099–5102.

Solter, D., B.B. Knowles (1978) Monoclonal antibody defining a stage-specific mouse embryonic antigen (SSEA-1). Proc Natl Acad Sci USA *75*: 5565–5569.

Stewart, C.L. (1991) Prospects for the establishment of embryonic stem cells and genetic manipulation of domestic animals; in Pedersen, R.A., A. McLaren, N.L. First (eds): Animal Applications of Research in Mammalian Development. Cold Spring Harbor, Cold Spring Harbor Laboratory Press, pp 267–283.

Stewart, C.L., I. Gadi, H. Bhatt (1994) Stem cells from primordial germ cells can reenter the germ line. Dev Biol *161*: 626–628.

Stice, S.L., J.M. Robl, F.A. Ponce de Leon, J. Jerry, P.G. Golueke, J.B. Cibelli, J.J. Kane (1998) Cloning: New breakthroughs leading to commercial opportunities. Theriogenology *48*: 129–138.

Stice, S.L., N.S. Strelchenko, C.L. Keefer, L. Matthews (1996) Pluripotent bovine embryonic cell lines direct development following nuclear transfer. Biol Reprod *54*: 100–110.

Strelchenko, N. (1996) Bovine pluripotent stem cells. Theriogenology *45*: 131–140.

Strelchenko, N., J. Betthauser, G. Jurgella, E. Farsberg, P. Damiani, P. Golueke (1998) Use of somatic cells in cloning (abstract). Proc Gen Engineering Cloning Anim, Park City.

Strojek, R.M., M.A. Reed, J.L. Hoover, T.E. Wagner (1990) A method for cultivation morphologically undifferentiated embryonic stem cells from porcine blastocysts. Theriogenology *33*: 901–913.

Strojek-Baunack, R.M., K. Bürkle, K. Burich, S. Hense, C. Reintjes, J. Hahn (1991) Kokultivierung von Rinderblastozysten unter verschiedenen Bedingungen in vitro. Fertilität *7*: 77–84.

Sukoyan, M.A., A.N. Golubitsa, A.I. Zhelezova, A.G. Shilov, S.Y. Vatolin, L.P. Maximovsky, L.E. Andreeva, J. McWhir, S.D. Pack, S.I. Bayborodin, A.Y. Kerkis, H.I. Kizilova, O.L. Seov (1992) Isolation and cultivation of blastocyst-derived stem cell lines from American mink (*Mustela vison*). Mol Reprod Dev *33*: 418–431.

Sukoyan, M.A., S.Y. Vatolin, A.N. Golubitsa, A.I. Zhelezova, L.A. Semenova, O.L. Serov (1993) Embryonic stem cells derived from morulae, inner cell mass, and blastocysts of mink: Comparisons of their pluripotencies. Mol Reprod Dev *36*: 148–158.

Sun, L., C.S. Bradford, G. Ghosh, P. Collodi (1995) ES-like cell cultures derived from early zebrafish embryos. Mol Mar Biol Biotechnol *4*: 193–199.

Takagi, Y., N.C. Talbot, C.E. Rexroad, V.G. Pursel (1997) Identification of pig primordial germ cells by immunocytochemistry and lectin binding. Mol Reprod Dev *46*: 567–580.

Talbot, N.C., A.M. Powell, C.E. Rexroad (1995) In vitro pluripotency of epiblast derived from bovine blastocysts. Mol Reprod Dev *42*: 35–52.

Talbot, N.C., C.E. Rexroad, V.G. Pursel, A.M. Powell (1993) Alkaline phosphatase staining of pig and sheep epiblast cells in culture. Mol Reprod Dev *36*: 139–147.

Thomas, K.R., M.R. Capecchi (1987) Site-directed mutagenesis by gene targeting in mouse embryo-derived stem cells. Cell *51*: 503–512.

Thomson, J.A., J. Itskovitz-Eldor, S.S. Shapiro, M.A. Waknitz, J.J. Swiergiel, V.S. Marshall, J.M. Jones (1998) Embryonic stem cell lines derived from human blastocysts. Science *282*: 1145–1147.

Thomson, J.A., J. Kalishman, T.G. Golos, M. Durning, C.P. Harris, R.A. Becker, J.P. Hearn (1995) Isolation of a primate embryonic stem cell line. Proc Natl Acad Sci USA *92*: 7844–7848.

Thomson, J.A., J. Kalishman, T.G. Golos, M. Durning, C.P. Harris, J.P. Hearn (1996) Pluripotent cell lines derived from common marmoset *(Callithrix jacchus)* blastocysts. Biol Reprod *55*: 254–259.

Thoraval, P., F. Lasserre, F. Coudert, G. Dambrine (1994) Somatic and germline chicken chimeras obtained from brown and white Leghorns by transfer of early blastodermal cells. Poul Sci *73*: 1897–1905.

Trounson, A., M. Pera (1998) Potential benefits of cell cloning for human medicine. Reprod Fertil Dev *10*: 121–125.

Van Stekelenburg-Hamers, A.E.P., T.A.E. Van Achterberg, H.G. Rebel, J.-E. Flechon, K.H.S. Campbell, S.M. Weima, C.L. Mummery (1995) Isolation and characterization of permanent cell lines from inner cell mass cells of bovine blastocysts. Mol Reprod Dev *40*: 444–454.

Vasiliev, I.M., S.G. Vasilieva (1995) Factors influencing the isolation of embryonic stem cells of pigs. Russ J Dev Biol *26*: 163–166.

Vasilieva, S.G., K. Prelle, S. Müller, U. Besenfelder, M. Müller, G. Brem (1998) Establishment and long-term culture of rabbit ES cells. Russ J Dev Biol *5*: 347–353.

Wakamatsu, Y., K. Ozato, T. Sasado (1994) Establishment of a pluripotent cell line derived from a medaka *(Oryzias latipes)* blastula embryo. Mol Mar Biol Biotechnol *3*: 185–191.

Wakayama, T., A.C.F. Perry, M. Zuccotti, K.R. Johnson, R. Yanagimachi (1998) Full-term development of mice from enucleated oocytes injected with cumulus cell nuclei. Nature *394*: 369–374.

Wang, Z.Q., F. Kiefer, P. Urbanek, E.F. Wagner (1997) Generation of completely embryonic stem cell-derived mutant mice using tetraploid blastocyst injection. Mech Dev *62*: 137–145.

Wells, D.N., P.M. Misica, T.A.M. Day, H.R. Tervit (1997) Production of cloned lambs from an established embryonic cell line: A comparison between in vivo- and in vitro-matured cytoplasts. Biol Reprod *57*: 385–393.

Wells, D.N., P.M. Misica, H.R. Tervit (1999) Production of cloned calves following nuclear transfer with cultured adult mural granulosa cells. Biol Reprod *60*: 996–1005.

Wheeler, M.B. (1994) Development and validation of swine embryonic stem cells: A review. Reprod Fertil Dev *6*: 563–568.

Wheeler, M.B., S.-J. Choi (1997) Embryonic stem cells and transgenics: Recent advances. Arq Fac Vet UFRGS *25(suppl)*: 64–83.

Wheeler, M.B., A.L. Rund, G.T. Bleck (1995) The use of embryonic stem cells in the production of transgenic livestock. Emb Trans News *13*: 20–25.

Wianny, F., C. Perreau, M.T. Hochereau-de Reviers (1997) Proliferation and differentiation of porcine inner cell mass and epiblast in vitro. Biol Reprod *57*: 756–764.

Wilmut, I., A.E. Schnieke, J. McWhir, A.J. Kind, K.H.S. Campbell (1997) Viable offspring derived from fetal and adult mammalian cells. Nature *385*: 810–813.

Wobus, A.M., K. Guan (1998) Embryonic stem cell-derived cardiac differentiation: Modulation of differentiation and 'loss-of-function' analysis in vitro. Trends Cardiovasc Med *8*: 64–74.

Wobus, A.M., H. Holzhausen, P. Jäkel, J. Schöneich (1984) Characterization of a pluripotent stem cell line derived from a mouse embryo. Exp Cell Res *152*: 212–219.

Wobus, A.M., G. Kaomei, J. Shan, M.-C. Wellner, J. Rohwedel, J. Guanju, B. Fleischmann, H.A. Katus, J. Hescheler, W.-M. Franz (1997a) Retinoic acid accelerates embryonic stem cell-derived cardiac differentiation and enhances development of ventricular cardiomyocytes. J Mol Cell Cardiol *29*: 1525–1529.

Wobus, A.M., J. Rohwedel, V. Maltsev, J. Hescheler (1994) In vitro differentiation of embryonic stem cells into cardiomyocytes or skeletal muscle cells is specifically modulated by retinoic acid. Rouxs Arch Dev Biol *204*:36–45.

Wobus, A.M., J. Rohwedel, C. Strübing, J. Shan, K. Adler, V. Maltsev, J. Hescheler (1997b) In vitro differentiation of embryonic stem cells; in Klug, S, R. Thiel (eds): Methods in Developmental Toxicology and Biology. Berlin, Blackwell Science, pp 1–17.

Wobus, A.M., G. Wallukat, J. Hescheler (1991) Pluripotent mouse embryonic stem cells are able to differentiate into cardiomyocytes expressing chronotropic responses to adrenergic and cholinergic agents and Ca^{2+} channel blockers. Differentiation *48*: 173–182.

Wolf, E., R. Kramer, I. Polejaeva, H. Thoenen, G. Brem (1994) Efficient generation of chimeric mice using embryonic stem cells after long-term culture in the presence of ciliary neurotrophic factor. Transgenic Res *3*: 152–158.

Wolf, E., R. Kramer, H. Thoenen, G. Brem (1996) Alternative culture systems for mouse embryonic stem cells; in Arnold, W., P. Köpf-Maier, B. Micheel (eds): Immunodeficient Animals: Models for Cancer Research. Contrib Oncol. Basel, Karger, pp 214–218.

Wolf, E., V. Zakhartchenko, G. Brem (1998) Nuclear transfer in mammals: Recent developments and future perspectives. J Biotechnol *65*: 99–110.

Wood, S.A., N.D. Allen, J. Rossant, A. Auerbach, A. Nagy (1993) Non-injection method for the production of embryonic stem cell-embryo chimeras. Nature *365*: 87–89.

Wylie, C.C. (1993) The biology of primordial germ cells. Eur Urol *23*: 62–67.

Zakhartchenko, V., G. Durcova-Hills, W. Schernthaner, M. Stojkovic, H.-D. Reichenbach, S. Müller, R. Steinborn, M. Müller, H. Wenigerkind, K. Prelle, E. Wolf, G. Brem (1999a) Potential of fetal germ cells for nuclear transfer in cattle. Mol Reprod Dev *52*: 421–426.

Zakhartchenko, V., G. Durcova-Hills, M. Stojkovic, W. Schernthaner, K. Prelle, R. Steinborn, M. Müller, G. Brem, E. Wolf (1999b) Effects of serum starvation and recloning on the efficiency of nuclear transfer using bovine fetal fibroblasts. J Reprod Fertil *115*: 325–331.

Cells Tissues Organs

Cells Tissues Organs 1999;165:237–245

Accepted after revision: May 30, 1999

Can Exogenous Stem Cells Be Used in Transplantation?

Kenneth R. Boheler[a] Marc Y. Fiszman[b]

[a]National Institute on Aging, Intramural Research Program, Gerontology Research Center, Laboratory of Cardiovascular Science, Baltimore, Md., USA; [b]INSERM U523, Institute of Myology, Hôpital Pitié-Salpêtrière, Paris, France

Key Words

Embryonic stem cells · Cell replacement therapy

Abstract

Today's most urgent problem in transplantation is the lack of suitable donor organs and tissues and as the population ages, demands for organs and tissue therapies will only increase. One alternative to organ transplantation is cell therapy whose aim is to replace, repair or enhance the biological function of damaged tissue or diseased organs. One goal of cellular transplantation thus has been to find a renewable source of cells that could be used in humans. Embryonic stem (ES) cells have the potential to proliferate in vitro in an undifferentiated and pluripotent state. Theoretically, ES cells are capable of unlimited proliferation in vitro. ES cells spontaneously differentiate into derivatives of all three primary germ layers: endoderm, ectoderm and mesoderm, hence providing cells in vitro which can theoretically be isolated and used for transplantation. Furthermore, these pluripotent stem cells can potentially be used to produce large numbers of cells that can be genetically modified in vitro. Once available, this source of cells may obviate some of the critical needs for organ transplantation. Murine ES cells have been extensively studied and all available evidence indicates that all aforementioned expectations are indeed fulfilled by ES cells. ES cells as well as embryonic germ cells have recently been isolated and maintained in culture. The recent descriptions of human ES cells portend the eventual use of allogeneic in vitro differentiated cells for human therapy. This goal, however, is fraught with obstacles. Our aim is first to review the recent advances made with murine ES cells and then to point out potentials and difficulties associated with the use of human ES cells for transplantation.

Abbreviations used in this paper

EG	embryonic germ cells
ES	embryonic stem cells
SCID	severe combined immune deficiency
tTA	tetracycline transactivator

KARGER

Fax + 41 61 306 12 34
E-Mail karger@karger.ch
www.karger.com

1422–6405/99/1654–0237$17.50/0

Accessible online at:
www.karger.com/journals/cto

Dr. Marc Y. Fiszman, INSERM U523
Institute of Myology, Hôpital Pitié-Salpêtrière
47, Boulevard de l'Hôpital, F–75651 Paris (France)
Tel. +33 1 42 16 57 07, Fax +33 1 42 16 57 00
E-Mail mfiszman@myologie.infobiologen.fr

Introduction

Quiescent stem cells or combined progenitor cells allow organ or tissue self-repair; however, tissue damage or disease that disrupts cell function requires therapy in the form of transplantation or cell replacement. Today's most urgent problem in transplantation is the lack of suitable donor organs and tissues. In fact as few as 5% of the organs needed in the US will be available for those awaiting transplantation [Platt, 1998]. As the population ages, demands for organs and tissue therapies will only increase. One alternative to organ transplantation is cell therapy whose aim it is to replace, repair or enhance the biological function of damaged tissue or diseased organs [Gage, 1998]. One goal of cellular transplantation thus has been to find a renewable source of cells that could be used in humans.

Such an approach is potentially achievable by the transfer of isolated and characterized cells to a target organ in sufficient number and quality for them to survive and restore function. Sources of cells or tissues are self (autologous), same species (allogeneic), different species (xenographic), primary or immortalized cells (cell lines) and differentiated stem cells. The ability to cultivate, multiply and manipulate these cell types has either limited or encouraged their use in specific treatment protocols [Gage, 1998]. Where possible, autologous cells are preferential but frequently unavailable. The recent descriptions of human embryonic stem (ES) cells portend the eventual use of allogeneic in vitro-differentiated cells for human therapy. This goal is fraught with obstacles and will require integration of cell biology, immunology, tissue engineering, molecular biology, materials science, transplantation biology and the clinical expertise relevant to the disease being treated for any eventual intervention.

Recently an important first step towards reaching this goal has been accomplished by two laboratories, one of which has described the establishment of human ES cells and the other human embryonic germ (EG) cells [Shamblott et al., 1998; Thomson et al., 1998]. As cells in vitro can theoretically be isolated and used for transplantation, pluripotent stem cells can potentially be used to produce large numbers of cells that can be genetically modified in vitro. Once available, this source of cells may obviate some of the critical needs for organ transplantation. In order to appreciate the potential importance of these human ES or EG cells, it is worth examining the recent advances made with mouse ES cells.

Mouse ES Cells: Properties and Usefulness

Mouse ES cells have originally been derived from undifferentiated cells of the inner cell mass of preimplantation blastocysts or from eight-cell embryos [Doetschman et al., 1985, 1988]. These cells can proliferate in vitro in an undifferentiated, pluripotent state on a feeder layer of primary mouse embryonic fibroblasts or leukemia inhibitory factor-producing mouse fibroblastic cell lines (STO cells) [Williams et al., 1988]. Theoretically, ES cells are capable of unlimited proliferation in vitro. When injected into mice, ES cells can form teratomas that contain cell types derived from all three germ layers. Importantly, these cells in vitro can maintain a relatively normal and stable karyotype even with continual passaging. Upon removal of the mouse embryonic fibroblast feeder layer or more generally in the absence of leukemia inhibitory factor, ES cells spontaneously differentiate into derivatives of all three primary germ layers: endoderm, ectoderm and mesoderm [Evans and Kaufman, 1981; Martin, 1981; Wobus et al., 1984]. The in vitro differentiation of ES cells requires their initial aggregation to form structures, termed embryoid bodies, which then differentiate into a variety of specialized cell types, including cardiomyocytes (atrial, ventricular, pacemaker), smooth and skeletal muscle cells, hematopoietic cells, cartilage, melanocytes and neurons.

When returned to the embryonic environment by transfer into a host blastocyst or aggregation with blastomere stage embryos, ES cells behave like normal embryonic cells. They can contribute to all tissues in the resulting chimeras including gametes. ES cells thus have the full potential to develop along all lineages of the embryo proper. Importantly, genetic alterations introduced into ES cells can also be transmitted to the germline through the production of ES cell chimeras. Applying gene targeting technology to ES cells in culture allows genetic modification of endogenous genes and study of their function in vivo.

All these general properties explain the wide variety of applications for which ES cells have been instrumental.

Creation of New Animal Models

The existence of ES cells has allowed the creation, almost at will, of any kind of mutation in any given gene through the use of a wide variety of techniques which include selective homologous recombination and exon or promoter trap methods. So far, most animal models have been created as substitutes for human pathologies, and the use of ES cell technology in this context has only been

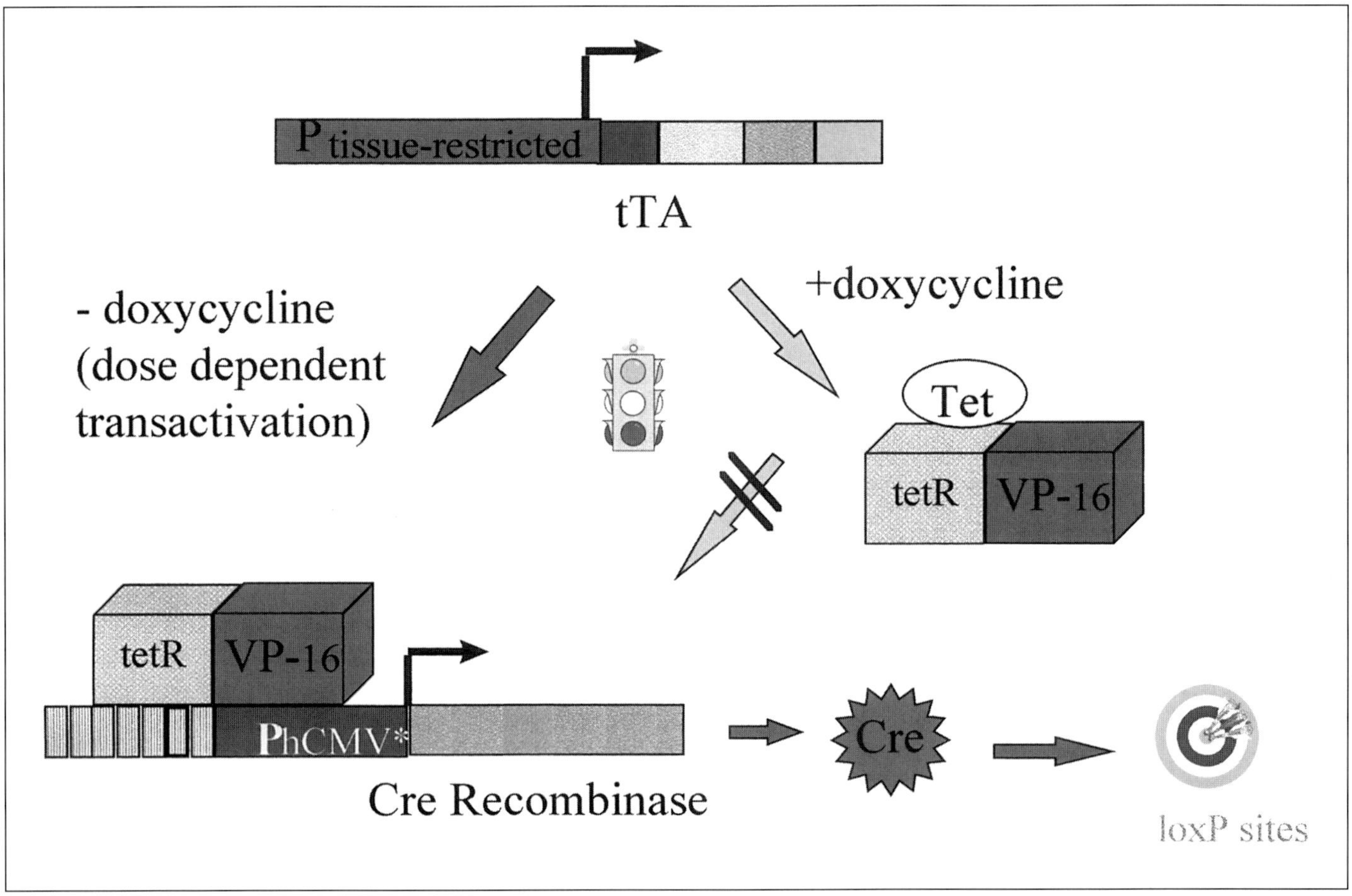

Fig. 1. Binary system for inducible Cre recombinase gene expression. In the binary system, tissue specificity is generated through use of a tissue-restricted promoter that regulates tTA gene transcription. The target for tTA is a minimal promoter fused downstream of seven operator sequences (TRE) of the *Escherichia coli* tet operon. Being bacterially derived, this gene is inactive in mammalian cells, unless the tTA protein is expressed in the same cell. Temporal control over Cre recombinase transgene expresssion is achieved in the presence or absence of doxycycline. Use of the tTA system thus permits the study of gene expression in specific ES-derived cells or mouse tissues during specific developmental stages.

successfully applied to mice. One such example is provided by the mouse model in which the cardiac myosin heavy chain has been mutated ($Arg^{403} \rightarrow Gln$) in order to reproduce familial hypertrophic cardiomyopathy [Geisterfer-Lowrance et al., 1996]. Many of these models were somewhat disappointing mostly because pathological manifestations are the end result of an array of secondary effects which may be different between humans and mice.

Another type of animal model using ES cells has allowed the creation of animals in which large-scale chromosome alterations have been engineered. Two such models were recently described [Smith et al., 1995; Van Deursen et al., 1995] which take advantage of the Cre-loxP system. Very briefly, loxP sites are introduced at specific sites in the genome, on different chromosomes, and the cells, that carry these sites, are further transfected by a Cre recombinase-carrying vector. As a very rare event, the Cre recombinase will induce a recombination between the two sites, hence leading to a chromosomal translocation. Smith and his colleagues targeted the translocation between the myc gene (chromosome 15) and immunoglobulin heavy chain genes (chromosome 12) while Van Deursen and his colleagues obtained a translocation between the Dek gene (chromosome 13) and the Can gene (chromosome 2). Such a strategy should allow the design of a variety of chromosome rearrangements specific to different types of cancer or of various pathologies.

Cre recombinase and loxP sites can also be used in a binary system to temporally and/or spatially regulate gene expression. Although several inducible systems exist [Kuhn et al., 1995; Logie and Stewart, 1995], perhaps the best regulatory system of this sort relies on tetracycline regulation as developed by Bujard [Gossen and Bujard, 1992]. When placed under the control of a lineage-specific promoter (e.g., α-myosin heavy chain promoter in heart) tissue-restricted expression of the tetracycline transactivator (tTA) can be achieved [Yu et al., 1996; Redfern et al., 1999]. In this system, doxycycline administered in the mouse diet binds to the tTA, preventing activation of the tetO promoter, whereas in the reverse tTA system, transcriptional activation is induced by addition of doxycycline. In the binary system, inducible control of Cre recombinase would be regulated by binding of the tTA to the tetO promoter (fig. 1). Under conditions where the tetO promoter is not activated, sequences containing the loxP site should remain intact; however, tTA transactivation by withdrawal of doxycycline would be expected to bind to the tetO promoter and drive expression of the Cre recombinase. Available loxP sites would then be targeted. The advantage of such a system rests in its ability to induce the excision of a critical exon in the target gene flanked by loxP sites at specific developmental stages (embryonic, fetal, perinatal adult). Such a system has already been proven to be functional in mice [Mayford et al., 1996; Tsien et al., 1996].

In vitro Studies

Approaches such as those described for making animal models require prior knowledge of the sequence and genomic organization of each gene to be studied. The development of ES cells and their study in vitro has, however, provided a novel approach to study development or study the effects of mutations in the genome before their introduction into the mouse germline. For example, differentiation of ES cells and formation of embryoid bodies have provided a major opportunity to study the regulation of lineage commitment. This approach has led to the identification of novel precursor cells [Choi et al., 1998] and has allowed a better characterization of specific pathways such as the expression of myogenic factors during skeletal muscle differentiation [Braun and Arnold, 1996]. Additionally, ES cells have been used as a complementary system to study promoter elements instead of using transgenic animals [Pari et al., 1991]. For example, the random insertion of exogenous DNA into single sites in the mammalian genome (gene trapping) provides a genome-wide strategy for functional genomics. When performed in tandem with the developmental potential of ES cells to differentiate into distinct cell lineages (expression trapping), such techniques are useful for the identification of novel genes expressed in developing systems. Stanford et al. [1998] successfully used this system to identify several novel hematopoietic and vascular genes.

ES cells have also been used to study the function of genes. In particular the system is well suited to understand the role of genes through 'loss of function' studies by developing ES cell lines that lack a specific gene. This is of particular interest when the inactivation of the gene is lethal in the animal. Such an example was recently published in the case of ES cells lacking the β_1 integrin [Rohwedel et al., 1998; Wobus and Guan, 1998]. Recently, we successfully prepared several neomycin-resistant R1 ES cell clones containing exons for the cardiac calcium release channel flanked by loxP sites. Knockout of this gene results in embryonic lethality, but in the context of this recombination event, flanking of the exon by loxP sites should not have affected gene expression. To ensure that these ES cell clones acted similarly to the parental cells before injection into mouse blastocysts, the undifferentiated ES cell clones were differentiated in vitro [Maltsev et al., 1994]. Clonal ES cells were trypsinized, and hanging drops were prepared with 600 or 800 ES cells/20 μl of DMEM media containing 20% FBS and additives. Following 2 days in a hanging drop, the developing embryoid body was transferred to suspension culture with maintenance media (DMEM/15% FBS/additives). After 4–5 more days in suspension culture, the embryoid bodies were plated and allowed to attach. Within 1–5 days, >90% of the plated embryoid bodies from both the parental and clonal ES cell lines contained spontaneously and rhythmically contracting cardiomyocytes (unpubl. results). The results suggested that modifications in the introns of this gene had not adversely affected its expression after homologous recombination. Conversely, ES cells have also been an ideal in vitro system for 'gain of function' studies, since they permit the study of specific gene overexpression on a given differentiation lineage without having to worry about the effects of such an overexpression on the overall embryonic development [Grépin et al., 1997].

Cell Therapies

Well-defined differentiation protocols from mouse ES cells have been developed for neuronal cells [Gottlieb and Huettner, 1999], adipocytes [Dani, 1999], smooth muscle myocytes [Drab et al., 1997], skeletal muscle myocytes and hematopoietic cells [Wiles and Keller, 1991]. A number of protocols have also been described for the production and isolation of cardiomyocytes [Wobus et al., 1991]. Spontaneously and rhythmically contracting cardiomyocytes for example can be isolated by mechanical dissection and collagenase treatment. Individual cells can then be used in single cell assays. ES-derived cardiomyocytes express α- and β-myosin heavy chain, α-tropomyosin, myosin light chain-2v, atrial natriuretic factor, phospholamban, L-type calcium channels [Hescheler et al., 1997] SR CaATPase (unpubl. data), ryanodine receptor type 2 (unpubl. data), and they show normal contractile sensitivity to calcium. Electrophysiological analyses have identified action potentials typical for atrial, ventricular and conduction system cardiomyocytes. The cells in vitro do not, however, seem to form T tubules or show several other characteristics of adult cardiomyocytes.

Even when optimal conditions are used to induce differentiation, the percentage of differentiated cells belonging to one phenotype still remains very small (in the case of cardiac differentiation, this percentage is frequently of the order of 3–5%). Therefore, one of the major problems which arises is: how can one isolate a pure population of ES cell-derived differentiated cells that could potentially be used for transplantation?

Several groups have been successful in isolating individual cell types following genetic manipulation and in vitro differentiation of ES cells. Dinsmore et al. [1996] reported the differentiation of ES cells into nearly homogeneous populations of neurons and skeletal muscle cells. For skeletal muscle cell isolation, ES cells were stably transfected with a gene coding for the muscle-specific regulatory factor MyoD. When induced with DMSO, the ES cells expressing high levels of MyoD differentiated into skeletal myoblasts that fused and formed myotubes capable of contraction. Greater than 90% of those cells expressing high levels of MyoD were found to be of skeletal muscle origin [Dekel et al., 1992]. Li et al. [1998] described a system which allowed for the efficient purification of neuroepithelial progenitor cells. Specifically, they targeted a promoterless neomycin cassette to the SOX2 gene, so that only those cells expressing SOX2 following in vitro differentiation would be neomycin (G418)-resistant. The SOX2 gene encodes for a DNA binding protein whose expression is restricted to the early neural plate. The neomycin-resistant cells were shown to differentiate into neuronal networks in the absence of other cell types.

Field and his group have successfully shown that genetically selected cardiomyocytes can be isolated from differentiating ES cells and could thereafter be used to form stable intracardiac grafts [Klug et al., 1996]. This group prepared an ES cell line carrying a fusion gene comprised of the cardiac α-myosin heavy chain promoter and sequences encoding aminoglycoside phosphotransferase to confer neomycin (G418) resistance. Following stable transfection, they were able to select a relatively pure population of cardiomyocytes (99.6%) and inject them into the ventricular myocardium of mdx mice. Subsequent analyses indicated the presence of ES-derived cardiomyocyte grafts for at least 7 weeks after implantation.

Lastly it is also possible to isolate ES-derived differentiated cells without having to rely on any selection procedure but through the use of the specific expression of a fluorophore by the differentiated cells. Such an example was shown in the case of cardiac differentiation where ES cells were stably transfected by the gene encoding the green fluorescent protein under the control of the cardiac α-actin promoter [Kolossov et al., 1998]. Upon differentiation only cardiomyocytes express the green fluorescent protein and it is possible to isolate them by flow-cytometric methods.

All these data show that specific and relatively pure cell types can be isolated from in vitro differentiated ES cells. Whether similar success will be forthcoming from the human ES cells is unknown, but the prospects seem very favorable.

Human ES Cells: ES and EG Cells

In November 1998, Thomson et al. [1998] reported the derivation of the first ES cell lines from human blastocysts. Several characteristics of ES cells were determined including a high ratio of nucleus to cytoplasm, prominent nucleoli, and colony morphology similar to that of ES cells derived from rhesus monkey (table 1). One of the ES cell lines isolated retained a normal XX karyotype after 6 months of culture. This human ES cell line expressed high levels of telomerase activity and expressed cell surface markers characteristic of undifferentiated nonhuman primate ES and human embryonic carcinoma cells. The undifferentiated cells showed positive alkaline phosphatase activity. Upon injection into severe combined immunodeficient (SCID) beige mice, teratomas were formed and derivatives of all three germ layers identified. With these tests, the evidence is very good that human ES cells have in fact been derived.

Shamblott et al. [1998] subsequently reported the derivation of pluripotent stem cells from cultured human primordial germ cells. Although examined to a lesser extent

Table 1. Characteristics of pluripotent embryonic cell lines: mouse versus human

	Mouse ES cells	Human ES cells	Human EG cells
Undifferentiated			
Unlimited, undifferentiated proliferation	Yes	Probable	Possible
Compact, multilayered colonies	Yes	Yes	Yes
High nuclear to cytoplasmic ratio	Yes	Yes	Yes
Alkaline phosphatase activity	Yes	Yes	Yes
High levels of telomerase activity	Yes	Yes	Not tested
Stable developmental potential	Yes	Yes	Possible
Differentiated			
Potential for in vitro differentiation	Yes	Yes	Yes
Express conserved epitope markers	Yes	Yes	Yes
Ability to give rise to multiple cell types	Yes	Yes	Yes
Ability to contribute to germline	Yes	Unknown/ unethical	Unknown/ unethical

than the ES cells described by Thomson et al., these cells also showed many of the characteristics of pluripotency: high levels of AP activity, compacted multicellular colonies, positive staining for a number of appropriate germline markers, and maintenance of normal karyotypic features. The immortality of these cells remains to be demonstrated.

Potential Developments for Human ES/EG Cells

The derivation of these stem cells represents the first major step in obtaining a renewable, tissue culture source of human cells capable of differentiating into a wide variety of cell types. The implications that relate to the development of these cells have broad applications for basic research and potential clinical use in humans. These pluripotent cells will be useful for in vitro studies of normal human embryogenesis, abnormal development (generation of cell lines with modified genes and chromosomes), human gene discovery, and drug and teratogen testing.

Use in Basic Research

Their acquisition also bodes well for basic science investigations. Particularly relevant will be the study of development and their use in vitro to understand fundamental biological problems. As already mentioned, the availability of mouse ES cells has generated an enormous amount of information on the developmental programs associated with lineage commitment, on gene function and regulation of gene expression. The same types of studies should now be possible with human ES cells.

Cell Replacement Therapies

Cultivation in vitro and isolation of specific cell types should lead to their use as a renewable source of cells for tissue transplantation, cell replacement, and gene therapies. The group of Field has proven this concept experimentally in mouse heart not only with modified ES cells as described above, but with C2C12 myoblast and fetal cardiomyocyte grafts [Koh et al., 1993, 1995]. Exploitation of the newly derived human ES cells may provide a renewable and potentially unlimited source for cell replacement in humans. This latter application could eventually preclude the direct use of fetal tisssue in transplantation therapies. Clinical targets might include neurodegenerative disorders, diabetes, spinal cord injury, and hematopoietic repopulation and myocyte grafting. However, without even thinking of using these stem cells for transplantation therapy in humans, an exciting potential that they offer is the possibility to create animal models allowing investigators to study the fate of human cells. A number of studies have shown that it is possible to graft human cells in SCID mice and thus, provided that stem cells are available, it could be possible to create an animal in which a tissue (or part of it) is human. Such examples have already been described in the case of bone marrow-derived cells [Greiner et al., 1998]. Animals of that sort should prove to be very instrumental in the analysis of both normal and pathological development. An example of the potential of this approach has recently been published in the case of the McCune-Albright syndrome [Bianco et al., 1998]. Since ES cells can generate a wide variety of precursor cells it should, in principle, be possible to generate mice carrying the corresponding human tissues. Furthermore, since the ES cells can be modified, the available precursor cells should be both normal and pathological.

Limitations and Problems

Before the transfer of any ES-derived cells to humans can be accomplished, a number of experimental obstacles must be overcome and theoretical questions addressed. Any cells used in transplantation would need to be autologous/immunologically masked, expandable in vitro, show appropriate integration of selection vectors into either targeted or inconsequential random genomic sites, and have homogeneous cell phenotypes. Pure populations of cell types would be a prerequisite since introduction of contaminating pluripotent ES cells during transplantation might give rise to teratomas. In the scenario where human ES-derived differentiated cells – say cardiomyocytes, neuronal cells, hematopoietic cells or vascular smooth muscle cells – could be isolated, appropriate steps would need to be taken to prevent rejection of the transplanted cells. A first step towards this goal would be the establishment of protocols to isolate pure cell populations similar to what has been done with mouse ES cells in culture (see above).

In transplantation scenarios, prevention of rejection is one of the paramount issues as all stromal-derived and hematopoietic-derived (except RBCs) cells express either HLA class I and/or class II molecules which can be recognized by specific lymphocytes in a foreign host. There are many regimens utilized to immunosuppress hosts prior to transplant to facilitate graft acceptance. They include immunosuppressive therapy (i.e. cyclosporin A, Cytoxan), systemic tolerance to foreign HLA and immunotherapy to destroy graft-reactive lymphocytes. Only animals or humans with similar MHC/HLA molecules could be donors for other hosts. Practically this would require determination of allogeneic compatibility. It is quite easy to define the MHC (rodents) or HLA (humans) backgrounds. There are PCR-based systems to define human class I and class II expression as well as many antibodies for flow cytometry and immunohistological analysis. In rodents, it is much easier as many rodent species have already been defined so it would be simple to predict MHC differences and predict reactivities. For ES cells derived from one human individual, all the HLA molecules would be clonal. As such, banks of ES cells with known HLA backgrounds could be established. When subjected to appropriate genetic manipulation, specific cell types could be isolated; however, the need to isolate multiple pure populations of cells with defined HLA molecules represents an enormous amount of work. Perhaps in concert with the binary system described above, it should be possible to identify loci in the human genome where specific loxP sites could be inserted without deleterious functional consequences. Once prepared, it should be possible to use these loxP sites in conjunction with Cre recombinase to insert almost any vector construct into the human genome, permitting a rapid procedure for selecting pure populations.

Would it be possible to prepare 'universal donor cells' derived from ES cells? Theoretically, it could be possible if all class I and class II molecules were eliminated from the cell surface of ES cells. This would require extensive gene targeting and early attempts would probably have to be directed against class I molecules. The consequences of such extensive gene targeting are, however, difficult to assess. Finally, it would be necessary to prevent karyotypic changes during passaging and preparation of ES-derived cells. This is no small undertaking for even though cells may show normal karyotypes, there is no assurance that consequential changes in the genome have not occurred. As with nuclear transfer technology in the production of Dolly, it was thought that her genome would carry the mutations that had accumulated prior to nuclear transfer. Dolly is, however, still alive and her offspring seem perfectly normal. Perhaps this bodes well for the grafting of genetically modified material that has been passaged in tissue culture for very long periods of time, but the potential for genetic mutations that may have serious consequences cannot be discounted. The increased incidence of the large offspring syndrome from the cattle and sheep industry indicates some of the potential dangers of epigenetic phenomena associated with the preparation of cells in vitro for transfer to blastocysts [Young et al., 1998]. Perhaps such scenarios will not be of great consequence with sufficiently differentiated and selected cells; however, other techniques or rapid screening procedures (e.g. use of chip technology) may be necessary to control the quality of the cells introduced into a transplantation host.

Conclusions

The ES cell technology has revolutionized modern biology and has provided us with an unprecedented opportunity to understand normal development and to gain new insights into the mechanisms that lead to pathological situations. Mouse ES cells have provided the means to study development, gene expression and create models to look at the effects of specifically modified genes on mouse phenotype. Spatial and temporal control of gene expression in the mouse genome should permit better elucidation of the exact consequences resulting from altered gene expression/function in the absence of developmental compensation (as seen with traditional knockout techniques). The development of human ES cells is a major step towards the possibility of applying this technology to the treatment of human diseases. Although it is unclear if human ES cells will be as experimentally robust as mouse ES cells, in the event that human ES cells can be genetically modified and specific cell types isolated, their use in transplantation studies may become reality. A renewable source of human cells is the goal; however, much remains to be accomplished before in vitro-differentiated human ES cells can be used therapeutically.

Acknowledgements

The authors would like to thank Dr. Huang Tian Yang (NIA/GRC/LCS) for sharing unpublished results, and Dr. Dennis Taub (NIA/GRC/LI) for invaluable discussions related to the immunological questions associated with potential use of human ES cells.

References

Bianco, P., S.A. Kuznetsov, M. Riminucci, L.W. Fisher, A.M. Spiegel, P.G. Robey (1998) Reproduction of human fibrous dysplasia of bone in immunocompromised mice by transplanted mosaics of normal and Gsa-mutated skeletal progenitor cells. J Clin Invest *101*: 1737–1744.

Braun, T., H.-H. Arnold (1996) Myf5 and myoD are activated in distinct mesenchymal stem cells and determine different skeletal muscle cell lineage. EMBO J *15*: 310–318.

Choi, K., M. Kennedy, A. Kazarov, J.C. Papadimitriou, G. Keller (1998) A common precursor for hematopoietic and endothelial cells. Development *125*: 725–732.

Dekel, L., Y. Magal, S. Pearson-White, C.P. Emerson, M. Shani (1992) Conditional conversion of ES cells to skeletal muscle by an exogenous myoD1 gene. New Biol *4*: 217–224.

Dani, C. (1999) Embryonic stem cell-derived adipogenesis. Cells Tissues Organs *165*: 173–180.

Dinsmore, J., J. Ratliff, T. Deacon, P. Pakzaban, V. Jacoby, W. Galpern, V. Isacson (1996) Embryonic stem cells differentiated in vitro as a novel source of cells for transplantation. Cell Transplant *5*: 131–143.

Doetschman, T., P. Williams, N. Maeda (1988) Establishment of hamster blastocyst-derived embryonic stem (ES) cells. Dev Biol *127*: 224–227.

Doetschman, T.C., H. Eistetter, M. Katz, W. Schmidt, R. Kemler (1985) The in vitro development of blastocyst-derived embryonic stem cell lines: Formation of visceral yolk sac, blood islands and myocardium. J Embryol Exp Morphol *87*: 27–45.

Drab, M., H. Haller, R. Bychkov, B. Erdmann, C. Lindschau, H. Haase, I. Morano, F.C. Luft, A.M. Wobus (1997) From totipotent embryonic stem cells to spontaneously contracting smooth muscle cells: A retinoic acid and db-cAMP in vitro differentiation model. FASEB J *11*: 905–915.

Evans, M.J., M.H. Kaufman (1981) Establishment in culture of pluripotential cells from mouse embryos. Nature *292*: 154–156.

Gage, F.H. (1998) Cell therapy. Nature *392*: 18–24.

Geisterfer-Lowrance, A.A.T., M. Christe, D.A. Conner, J.S. Ingwall, F.J. Schoen, C.E. Seidman, J.G. Seidman (1996) A mouse model of familial hypertrophic cardiomyopathy. Science *272*: 731–734.

Gossen, M., H. Bujard (1992) Tight control of gene expression in mammalian cells by tetracycline-responsive promoters. Proc Natl Acad Sci USA *89*: 5547–5551.

Gottlieb, D.I., J.E. Huettner (1999) An in vitro pathway from embryonic stem cells to neurons and glia. Cells Tissues Organs *165*: 165–172.

Greiner, D.L., R.A. Hesselton, L.D. Shultz (1998) SCID mouse models of human stem cell engraftment. Stem Cells *16*: 166–177.

Grépin, C., G. Nemer, M. Nemer (1997) Enhanced cardiogenesis in embryonic stem cells overexpressing the GATA-4 transcription factor. Development *124*: 2387–2395.

Hescheler, J., B.K. Fleischmann, S. Lentini, V.A. Maltsev, J. Rohwedel, A.M. Wobus, K. Addicks (1997) Embryonic stem cells: A model to study structural and functional properties in cardiomyogenesis. Cardiovasc Res *36*: 149–162.

Klug, M.G., M.H. Soonpaa, G.Y. Koh, L.J. Field (1996) Genetically selected cardiomyocytes from differentiating embryonic stem cells form stable intracardiac grafts. J Clin Invest *98*: 216–224.

Koh, G.Y., M.G. Klug, M.H. Soonpaa, L.J. Field (1993) Differentiation and long-term survival of C2C12 myoblast grafts in heart. J Clin Invest *92*: 1548–1554.

Koh, G.Y., M.H. Soonpaa, M.G. Klug, H.P. Pride, B.J. Cooper, D.P. Zipes, L.J. Field (1995) Stable fetal cardiomyocyte grafts in the hearts of dystrophic mice and dogs. J Clin Invest *96*: 2034–2042.

Kolossov, E., B.K. Fleischmann, Q. Liu, W. Bloch, S. Viatchenko-Karpinski, O. Manzke, G.J. Li, H. Bohlen, K. Addicks, J. Hescheler (1998) Functional characteristics of ES cell-derived cardiac precursor cells identified by tissue-specific expression of the green fluorescent protein. J Cell Biol *143*: 2045–2056.

Kuhn, R., F. Schwenk, M. Aguet, K. Rjewsky (1995) Inducible gene targeting in mice. Science *269*: 1427–1429.

Li, M., L. Pevny, R. Lovell-Badge, A. Smith (1998) Generation of purified neural precursors from embryonic stem cells by lineage selection. Curr Biol *8*: 971–974.

Logie, C., A.F. Stewart (1995) Ligand-regulated site-specific recombination. Proc Natl Acad Sci USA *92*: 5940–5944.

Maltsev, V.A., A.M. Wobus, J. Rohwedel, M. Bader, J. Hescheler (1994) Cardiomyocytes differentiated in vitro from embryonic stem cells developmentally express cardiac-specific genes and ionic currents. Circ Res *75*: 233–244.

Martin, G.R. (1981) Isolation of a pluripotent cell line from early mouse embryos cultured in medium conditioned by teratocarcinoma stem cells. Proc Natl Acad Sci USA *78*: 7634–7638.

Mayford, M., M.E. Bach, Y.-Y. Huang, L. Lang, R.D. Hawkins, E.R. Kandel (1996) Control of memory formation through regulated expression of a CaMKII transgene. Science *274*: 1678–1683.

Pari, G., K. Jardine, M.W. McBurney (1991) Multiple CArG boxes in the human cardiac actin gene promoter required for expression in embryonic cardiac muscle cells developing in vitro from embryonal carcinoma cells. Mol Cell Biol *11*: 4796–4803.

Platt, J.L. (1998) New directions for organ transplantation. Nature *392*: 11–17.

Redfern, C.H., P. Coward, M.Y. Degtyarev, E.K. Lee, A.T. Kwa, L. Hennighausen, H. Bujard, G.I. Fishman, B.R. Conklin (1999) Conditional expression and signaling of a specifically designed Gi-coupled receptor in transgenic mice. Nat Biotechnol *17*: 165–169.

Rohwedel, J., K. Guan, W. Zuschratter, S. Jin, G. Ahnert-Hilger, D. Fürst, R. Fässler, A.M. Wobus (1998) Loss of β1 integrin function results in a retardation of myogenic but an acceleration of neuronal differentiation of embryonic stem cells in vitro. Dev Biol *201*: 167–184.

Shamblott, M.J., J Axelman, S. Wang, E.M. Bugg, J.W. Littlefield, P.J. Donovan, P.D. Blumenthal, G.R. Huggins, J.D. Gearhart (1998) Derivation of pluripotent stem cells from cultured human primordial germ cells. Proc Natl Acad Sci USA *95*: 13726–13731.

Smith, A.J.H., M.A.D. Sousa, B. Kwabi-Addo, A. Heppell-Parton, H. Impey, P. Rabbitts (1995) A site-directed chromosomal translocation induced in embryonic stem cells by Cre-loxP recombination. Nat Genet *9*: 376–385.

Stanford, W.L., G. Caruana, K.A. Vallis, M. Inamdar, M. Hidaka, V.O. Bautch, A. Bernstein (1998) Expression trapping: Identification of novel genes expressed in hematopoietic and endothelial lineages by gene trapping in ES cells. Blood *92*: 4622–4631.

Thomson, J.A., J. Itskovitz-Eldor, S.S. Shapiro, M.A. Waknitz, J.J. Swiergiel, V.S. Marshall, J.M. Jones (1998) Embryonic stem cell lines derived from human blastocysts. Science *282*: 1145–1147.

Tsien J.Z., D.F. Chen, D. Gerber, C. Tom, E.H. Mercer, D.J. Anderson, M. Mayford, E.R. Kandel, S. Tonegawa (1996) Subregion- and cell type-restricted gene knockout in mouse brain. Cell *87*: 1317–1326.

Van Deursen, J., M. Fornerod, B. Van Rees, G. Grosveld (1995) Cre-mediated site-specific translocation between nonhomologous mouse chromosome. Proc Natl Acad Sci USA *92*: 7375–7380.

Wiles, M.V., G. Keller (1991) Multiple hematopoietic lineages develop from embryonic stem (ES) cells in culture. Development *111*: 259–267.

Williams, R.L., D.J. Hilton, S. Pease, T.A. Willson, C.L. Stewart, D.P. Gearing, E.F. Wagner, D. Metcalf, N.A. Nicola, N.M. Gough (1988) Myeloid leukemia inhibitory factor maintains the developmental potential of embryonic stem cells. Nature *336*: 684–687.

Wobus, A.M., K. Guan (1998) Embryonic stem cell-derived cardiac differentiation: Modulation of differentiation and 'loss-of-function' analysis in vitro. Trends Cardiovasc Med *8*: 64–74.

Wobus, A.M., H. Holzhausen, P. Jäkel, J. Schöneich (1984) Characterization of a pluripotent stem cell line derived from a mouse embryo. Exp Cell Res *152*: 212–219.

Wobus, A.M., G. Wallukat, J. Hescheler (1991) Pluripotent mouse embryonic stem cells are able to differentiate into cardiomyocytes expressing chronotropic responses to adrenergic and cholinergic agents and Ca^{2+} channel blockers. Differentiation *48*: 173–182.

Young, L.E., K.D. Sinclair, I. Vilmut (1998) Large offspring syndrome in cattle and sheep. Rev Reprod *3*: 155–163.

Yu, Z., C.S. Redfern, G.I. Fishman (1996) Conditional transgene expression in the heart. Circ Res *79*: 691–697.

Editorial

Cells Tissues Organs

Cells Tissues Organs 1999;165:246–249

Embryonic Stem Cells: An Exciting Field for Basic Research and Tissue Engineering, but also an Ethical Dilemma?

H.-W. Denker

Institut für Anatomie, Universitätsklinikum Essen, Germany

Key Words
Embryonic stem cells · Ethics

Recent reports describing that, not unexpectedly, human embryonic stem (ES) cells have become available [Shamblott et al., 1998; Thomson et al., 1998] suddenly put ES cells in the centre of public interest. Mouse ES cells have been known as a fascinating object of biological research for many years, and in particular they have proven useful as an experimental tool for basic studies on gene function. Their potential use as model systems for studies on basic aspects of differentiation processes was in the past exploited by only a limited, but steadily increasing, number of researchers. Since last year interest in this field seems to have experienced a quantum leap, in particular due to the fact that the primate, and especially human, ES cells now available may show considerable promise for broad applications in tissue engineering and transplantation. Indeed, they could be an ideal source of material for transplantation purposes in the human. We will of course never have enough donated human cells and organs to meet all transplantation needs, in spite of all efforts that are being made. Xenotransplantation poses immunologic and animal protection problems. ES cells are attractive because they can proliferate and then be influenced by various cytokines to differentiate into certain directions, e.g. into the direction of haematopoietic stem cells that are of utmost interest in transplantation medicine [for animal experiments see e.g. Forrester et al., 1991; Palacios et al., 1995]. Studies along those lines are now being continued with human ES cells. ES cells are, on the other hand, also of considerable interest to embryologists and developmental biologists because they offer opportunities previously unavailable for performing studies on differentiation and pattern formation in mammals including (nonhuman) primates. It seems that the great practical use that derivatives of ES cell lines may find in transplantation medicine will increase the pressure to put much money and effort into such basic research on the regulation of differentiation in these systems.

I am very happy, therefore, that Anna Wobus and Kenneth Boheler have accepted to compile and edit this timely special issue of *Cells Tissues Organs*. I am sure it will be welcomed by the scientific community as a very impor-

KARGER

Fax +41 61 306 12 34
E-Mail karger@karger.ch
www.karger.com

1422–6405/99/1654–0246$17.50/0

Accessible online at:
www.karger.com/journals/cto

Prof. Dr. med. Dr. rer. nat. H.-W. Denker
Institut für Anatomie, Universitätsklinikum Essen
Hufelandstrasse 55
D–45122 Essen (Germany)
Tel. +49 201 723 4380/1, Fax +49 201 723 5916, E-Mail denker@uni-essen.de

tant actual source of information and will contribute to stimulating further discussions and further research in this field.

In all our excitement about ES cells we must, however, not overlook that, like many other technical achievements, these cells may also have a Janus-faced character. In particular, human ES cells confront us with a dilemma: as detailed below, primate ES cells may have the potential to form in vitro embryo-like structures and perhaps even complete embryos, which could raise grave ethical concerns. The question arises whether this could prohibit the use of these cells.

Let me make clear why I feel this latter, problematical aspect deserves our attention and why I think it asks for clarification by specifically designed experimental investigation. Non-human primate ES cell lines have been produced in the rhesus monkey and the marmoset *(Callithrix jacchus)* [Thomson et al., 1995, 1996]. Particularly remarkable are features of differentiation reported for the marmoset cell line. When propagated on a feeder cell layer as usual for ES cell culture, the cells can be maintained in an undifferentiated state. However, when these cells are allowed to differentiate by simply letting them grow at high density they develop embryo-like structures in vitro. These appear to be much more embryo-like than the 'embryoid bodies' known for a long time from work on mouse ES and teratocarcinoma cell lines (see below). Thomson et al. [1996] have published a picture (fig. 5) of such a spontaneously developed early embryo-like structure that appears to be identical with a normal primate embryo as found in vivo. It shows a well-formed embryonic shield, with amnion and an amniotic cavity, as well as a yolk sac. An early human embryo would look practically the same. Evidence for the differentiation of trophoblast, the cell type that normally forms the fetal part of the placenta, was also found in the marmoset ES cell line so that apparently every cell type was produced here in vitro. Even more importantly, a primitive streak-like structure was also formed in the epiblast in vitro. The primitive streak is one of the first signs of the main (craniocaudal) body axis and, at the same time, is the structural equivalent for the formation of two new germ layers, the mesoderm and the definitive endoderm. The formation of a normal-looking primitive streak in these embryo-like structures in vitro shows on one hand that they may indeed have to be addressed as complete embryos. On the other hand, formation of the primitive streak is thought to mark the onset of individuality, since monocygotic twinning is not possible after the primitive streak stage has been reached. Therefore, the primitive streak has been of utmost importance in all discussions about cloning, abortion and experimentation with human embryos. Development of the embryo-like structures (or embryos) in the monkey cell lines has not been followed further in vitro, as far as can be seen in the scientific literature, but taking these published data into account there can be no doubt that development could have proceeded to even more advanced stages, and there seems to be a high probability that well-formed embryos (rather than teratomas, see below) can grow in vitro. Given the high proliferative capacity of ES cells, this could then result in large numbers of identical, cloned embryos.

This is the only report I am aware of which claims that such a high degree of order can develop spontaneously in an in vitro culture of ES cells, but the same authors point out that this was not an isolated phenomenon [Thomson et al., 1996; Thomson and Marshall, 1998]. The formation of such a high degree of structural order in vitro is not known from mouse ES cells, although differentiation of virtually any cell type can be obtained and can already be manipulated to impressive degrees (as discussed in various contributions to this special issue of *Cells Tissues Organs*). Even if mouse 'embryoid bodies' are generated in vitro, a higher structural order is typically not obtained, as known from older literature [Sherman and Solter, 1975; Robertson, 1987]. At the organ level, an impressive degree of order has been observed after transplantation of primate ES cells ectopically into SCID mice, e.g. gut-like structures with not only a rather typical epithelium and villi, but even a rather typical stratification of the various layers of the gut wall including a lamina propria and various layers of musculature [Thomson and Marshall, 1998, fig. 9]. This is in line with many observations on spontaneously developing teratomas as well as on experimental teratomas in mice that have been well known for a long time [Sherman and Solter, 1975; Robertson, 1987]. What is not seen in spontaneously arising teratomas/teratocarcinomas or in teratomas produced after transplantation of embryos to ectopic sites is the formation of a harmonious whole embryo with intact main body axes. This has been seen, however, in a unique experiment with mouse ES cells transplanted into a temporary host embryo (which itself degenerated later because it consisted of lethal cells) [Nagy et al., 1993]. Until recently most people assumed that instructions from a host embryo are indeed essential here. The more remarkable appears to be the report by Thomson et al. [1996] that apparently a complete and well-shaped embryo with astonishingly well-developed extraembryonic membranes including an embryonic disc with primitive streak is formed in vitro by the marmoset ES cells. This sheds a special light on the well-documented observation that the environment into which totipotent early embryonic cells are brought greatly influences in particular axis formation and thus the formation of the general body plan. It must be remembered that even complete normal mouse embryos show disorganized differentiation and formation of teratomas when transplanted to ectopic sites, e.g. to the kidney [Sherman and Solter, 1975; Robertson, 1987].

In this context, we have to ask anew the question: what is so special about an egg or an early blastomere, and what do these cells provide specifically for the formation of a harmonious whole embryo? Are those 'fields of factors' or cytoplasmic determinants of the egg [if they exist in the mammalian system, as discussed recently again on the basis of new findings, Antczak and van Blerkom, 1997; Beddington and Robertson, 1999; Edwards and Beard, 1997; Gardner, 1997; Zernicka-Goetz, 1998; cf. Denker, 1976] dispensable in a confluent in vitro culture of primate ES cells? How is axis formation brought about and regulated under these conditions? These are very important questions that urgently need to be studied experimentally, and this can now be done with primate ES cells. I personally feel that it should, however, never be attempted with human ES cells, and the scientific desire to study the behaviour of human ES cells must be satisfied with analogies or homologies gleaned from comparative studies on non-human primate species.

Mammalian and human embryos are of course always thought to be necessarily derived from eggs. All discussions about the legal status of the embryo and about moral implications of embryo-related technology have this type of embryo in mind. Eggs are a rare type of cells, since only few egg cells are produced during the reproductive years of a woman. Even when follicle growth and egg cell production in the ovary are stimulated hormonally the number of eggs produced remains limited. (It may be added that our attitude seems to be different with respect to gametes that are normally produced in abundant quantity, i.e. sperm.) What also appears relevant to our traditional thinking and arguing is the limited capacity of the female genital tract which allows growth of and giving birth to only one or a few embryos at a time. I feel that the public attention that has been paid and is being paid for example to experiments on the cloning of farm animals like the sheep Dolly [Wilmut et al., 1997] using technologies based on egg-derived embryos must appear exaggerated in comparison with the potential that lies in ES cells, since all egg-based procedures still meet considerable technical problems, need a lot of time and effort and can therefore be done only on a very limited scale due to the limited availability of eggs. The situation would obviously be different if the same type of thing (i.e. cloning) could be done with ES cells alone that can be produced in unlimited quantity.

A major organizing element in an egg or embryo may be positional information, structuring cues, asymmetric distribution of molecules, and the way how these direct the formation of axes (dorsovertral, anterior-posterior, proximodistal). That this can theoretically be achieved by very simple physicochemical mechanisms was shown in computer models [Meinhardt, 1989, 1996]. Modern work about axis formation in the early insect embryo [St Johnston and Nüsslein-Volhard, 1992; Nilson and Schüpbach, 1998] identifies an increasing number of genes and gene products that appear to be the players in this game. The question to what extent these genes may be related to those that are at work in vertebrates including mammals and man, and how they might be related to any polar organization of the mammalian egg [see Antczak and van Blerkom, 1997; Gardner, 1997; Zernicka-Goetz, 1998] is naturally of great interest. Will ES cells be helpful here by providing an easily accessible experimental system? The remarkable ability of primate ES cells to form embryo-like structures (or even embryos) in vitro, mentioned above, may suggest this.

The specific feature of the egg can be thought to be that it provides very simple asymmetries and that it secures in this way a degree of structural order for the orthotopic formation of single body axes (in contrast to multiple or branched axes) so that the subsequent patterning of all tissues in the emerging embryo will lead to the formation of a harmonious, functional whole. It is a peculiar feature of eggs and totipotent blastomeres, on the other hand, that they can regulate defects and restitute such a harmonious whole. It is possibly due to this flexibility and to these mechanisms that, as we have seen in the example of ES cells in vitro, order can also develop spontaneously in a cell culture, in this case independent of an egg-derived prepattern. Here stochastic molecular events (of the type that can perhaps be best described by the chaos theory) may generate a first simple asymmetry. If eggs, early embryonic cells (blastomeres) and ES cells all have the same totipotency an exciting topic for ongoing and future research is of course whether they all make use of the same hierarchically dependent (and vectorial) processes of differentiation, cell movements and cell-to-cell interactions for the development of complex order.

An old aphorism, based on one of the founders of experimental embryology, Carl Ernst von Baer, sometimes mistaken as expressing preformistic views, states:

'Entwicklung ist Umbildung von einer Gestalt zur anderen' [as formulated by Seidel, 1960]. Will it have to be reformulated or re-interpreted in the light of the remarkable pattern-forming abilities of the ES cells? And what does all this mean for the ethical aspects and legal status of human ES cells?

Acknowledgement

I wish to thank PD Dr. med. Th. Heinemann, PD Dr. med. Ch. Viebahn and my wife, Dr. med. U. Denker for helpful comments.

References

Antczak, M., J. van Blerkom (1997) Oocyte influences on early development: The regulatory proteins leptin and STAT3 are polarized in mouse and human oocytes and differentially distributed within the cells of the preimplantation stage embryo. Mol Hum Reprod *3:* 1067–1086.

Beddington, R.S.P., E.J. Robertson (1999) Axis development and early asymmetry in mammals. Cell *96:* 195–209.

Denker, H.-W. (1976) Formation of the blastocyst: determination of trophoblast and embryonic knot; in Gropp, A., K. Benirschke (eds): Developmental Biology and Pathology. Current Topics in Pathology. Berlin, Springer, vol 62, pp 59–76.

Edwards, R.G., H.K. Beard (1997) Oocyte polarity and cell determination in early mammalian embryos. Mol Hum Reprod *3:* 863–905.

Forrester, L.M., A. Bernstein, J. Rossant, A. Nagy (1991) Long-term reconstitution of the mouse haematopoietic system by embryonic stem cell-derived fetal liver. Proc Natl Acad Sci USA *88:* 7514–7517.

Gardner, R. (1997) The early blastocyst is bilaterally symmetrical and its axis of symmetry is aligned with the animal-vegetal axis of the zygote in the mouse. Development *124:* 289–301.

Meinhardt, H. (1989) Models for positional signalling with aplication to the dorsoventral patterning of insects and segregation into different cell types. Development *107(suppl):* 169–180.

Meinhardt, H. (1996) Models of biological pattern formation: Common mechanism in plant and animal development. Int J Dev Biol *40:* 123–134.

Nagy, A., J. Rossant, R. Nagy, W. Abramow-Newerly, J.C. Roder (1993) Derivation of completely cell culture-derived mice from early-passage embryonic stem cells. Proc Natl Acad Sci USA *90:* 8424–8428.

Nilson, L.A., T. Schüpbach (1998) Localized requirements for windbeutel and pipe reveal a dorsoventral prepattern within the follicular epithelium of the Drosophila ovary. Cell *93:* 253–262.

Palacios, R., E. Golunski, J. Samaridis (1995) In vitro generation of hematopoietic stem cells from an embryonic stem cell line. Proc Natl Acad Sci USA *92:* 7530–7534.

Robertson E.J. (1987) Teratocarcinomas and Embryonic Stem Cells: A Practical Approach. Oxford, IRL Press.

Seidel, F. (1960) Körpergrundgestalt und Keimstruktur. Eine Erörterung über die Grundlagen der vergleichenden und experimentellen Embryologie und deren Gültigkeit bei phylogenetischen Überlegungen. Zool Anz *164:* 245–305.

Shamblott, M.J., J. Axelman, S. Wang, E.M. Bugg, J.W. Littlefield, P.J. Donovan, P.D. Blumenthal, G.R. Huggins, J.D. Gearhart (1998) Derivation of pluripotent stem cells from cultured human primordial germ cells. Proc Natl Acad Sci USA *95:* 13726–13731.

Sherman, M.I., D. Solter (1975) Teratomas and Differentiation. New York, Academic Press.

St Johnston, D., C. Nüsslein-Volhard (1992) The origin of pattern and polarity in the Drosophila embryo. Cell *68:* 201–219.

Thomson, J.A., J. Itskovitz-Eldor, S.S. Shapiro, M.A. Waknitz, J.J. Swiergiel, V.S. Marshall, J.M. Jones (1998) Embryonic stem cell lines derived from human blastocysts. Science *282:* 1145–1147.

Thomson, J.A., J. Kalishman, T.G. Golos, M. Durning, C.P. Harris, R.A. Becker, J.P. Hearn (1995) Isolation of a primate embryonic stem cell line. Proc Natl Acad Sci USA *92:* 7844–7848.

Thomson, J.A., J. Kalishman, T.G. Golos, M. Durning, C.P. Harris, J.P. Hearn (1996) Pluripotent cell lines derived from common marmoset *(Callithrix jacchus)* blastocysts. Biol Reprod *55:* 254–259.

Thomson, J.A., V.S. Marshall (1998) Primate embryonic stem cells. Curr Top Dev Biol *38:* 133–165.

Wilmut, I., A.E. Schnieke, J. McWhir, A.J. Kind, K.H.S. Campbell (1997) Viable offspring derived from fetal and adult mammalian cells. Nature *385:* 810–813.

Zernicka-Goetz, M. (1998) Fertile offspring derived from mammalian eggs lacking either animal or vegetal poles. Development *125:* 4803–4808.

Conference Calendar

2.12.–4.12.1999 **Berlin, Müggelsee** Germany	3rd Joint Meeting of the Society for Signal Transduction (STS) and the Signal Transduction Study Groups of the German Societies for Immunology (DGfI) and Cell Biology (DGZ)	**Information:** Internet www.sigtrans.de
11.12.–15.12.1999 **Washington, D.C.** USA	39th Annual Meeting of the American Society for Cell Biology	**Information:** The American Society for Cell Biology, 9650 Rockville Pike, Bethesda, MD 20814-3992 (USA) Tel. +1 301 530 7153, Fax +1 301 530 7139, E-Mail ascbinfo@ascb.org Internet www.ascb.org/ascb
8.1.–12.1.2000 **Maastricht** The Netherlands	26th Annual Conference of the International Embryo Transfer Society	**Information:** IETS, 1111 North Dunlap Avenue, Savoy, IL 61874 (USA) Tel. +1 (0) 217 356 3182, Fax +1 (0) 217 398 4119 E-Mail iets@assochq.org; Internet http://www.iets.uiuc.edu
27.1.–28.1.2000 **Halle (Saale)** Germany	3rd Symposium on Cytokines and Apoptosis in the Cardiovascular System	**Information:** PD Dr. H. Loppnow, Martin-Luther-Universität Halle-Wittenberg, Klinik und Poliklinik für Innere Medizin III, Forschungslabor, Magdeburgerstr. 21, D–06097 Halle (Saale) Tel. 0345 557 4541, Fax 0345 557 4542 E-Mail Harald.Loppnow@medizin.uni-halle.de
6.2.–11.2.2000 **Canberra** Australia	ACEM 2000, 16th Australian Conference on Electron Microscopy	**Information:** Dr. Stuart Craig, Fax +61 2 6246 5000, E-Mail craig@pi.csiro.au http://www.anu.au/EMU/acem
16.2.–20.2.2000 **Davos** Switzerland	Davos Tissue Engineering Workshop	**Information:** Prof. J. Hubbell Tel. +41 1 632 45 85, Fax +41 1 632 12 14
13.3.–16.3.2000 **Birmingham** UK	19th Joint Meeting of British Endocrine Societies	**Information:** British Endocrine Societies, 17/18 the Courtyard, Woodlands, Bradley Stoke, Bristol BS32 4NQ (UK) Tel. +44 1454 619036; +44 1454 616071 E-Mail info@endocrinology, org ; Internet http://www.endocrinology.org
26.3.–30.3.2000 **Karlsruhe** Germany	24. Jahrestagung der Deutschen Gesellschaft für Zellbiologie	**Information:** Prof. Dr. R. Paulsen, Universität Karlsruhe (T.H), Zoologisches Institut I, Zell- und Neurobiologie, Kornblumenstr. 13, D–76128 Karlsruhe (Germany) Tel. +49 721 608 2218, Fax +49 721 608 4848, E-Mail Reinhard.Pausen@bio-geo.uni-karlsruhe.de
30.3.–2.4.2000 **Dana Point, CA** USA	Serono Symposium: ART and the Human Blastocyst	**Information:** Cindy Bell, Serono Symposia USA, Inc., 100 Longwater Circle, Norwell, MA 02061 (USA) Tel. +1 781 681 2181, Fax +1 781 982 9481 E-Mail cindy.bell.us_bos01@serono.com
27.4.–29.4.2000 **Gothenburg** Sweden	3rd Symposium on Cartilage and Cartilage Repair in the New Millenium	Fax +41 31 934 10 16 E-Mail ecaservices@datacomm.ch
15.5.–20.5.2000 **Kamuela** Hawaii	World Biomaterials Congress	**Information:** Society for Biomaterials, 13355 Tenth Avenue North Suite 108, Minneapolis, MN 554241-5510 (USA) E-Mail registration@biomaterials.org

KARGER
Fax + 41 61 306 12 34
E-Mail karger@karger.ch
www.karger.com

Accessible online at:
www.karger.com/journals/cto

18.5.–20.5.2000 **Innsbruck** Austria	The International Tissue Engineering Symposium (ITEM 2000)	**Information:** Karin Klema, Laboratory for Tissue Engineering, Dept. of Oral and Maxillofacial Surgery, University Hospital – FKK, Maximilianstr. 10 A–6020 Innsbruck (Austria) Tel. +43 512 504/4373, Fax +43 512 504/4371 E-Mail maxillofaziale-chirurgie@uibk.ac.at
25.6.–28.6.2000 **Bologna** Italy	16th Annual Meeting of the European Society of Human Reproduction and Embryology	**Information:** ESHRE Central Office, Van Akenstraat 41, B–1850 Grimbergen (Belgium) Tel. +32 (0)2 269 09 69, Fax +32 (0) 2 269 56 00, E-Mail eshre@pophost.eunet.be
26.6.–30.6.2000 **Singapore**	7th Asia-Pacific Conference on Electron Microscopy	**Information:** Conf. Secretariat Fax +65 776 4971, E-Mail micngml@nus.edu.sg http://www.med.nus.edu.sg.micsoc/7apem
9.7.–14.7.2000 **Brno** Czech Republic	12th European Congress on Electron Microscopy	**Information:** EUREM 2000, Královopolská 147, CZ–61264 Brno (Czech Republic)
16.7.–20.7.2000 **Birmingham** UK	18th International Congress of Biochemistry and Molecular Biology beyond The Genome: Understanding and Exploiting Molecules and Cells in the Third Millennium	**Registration and further information:** 18th International Congress of Biochemistry and Molecular Biology, 59, Portland Place, London W1N 3AJ (UK) Tel. +44 171 580 5530, Fax +44 171-637-7626, E-Mail info@iubmb2000.org Internet http://www.iubmb2000.org
24.7.–26.7.2000 **Cambridge** UK	Tripartile Meeting, Anatomical Society of Great Britain and Ireland, Anatomische Gesellschaft und Nederlandse Anatomen Vereniging	**Information:** Dr. Jonathan Bennett, Programme Secretary, ASGBI, Division of Biometrical Sciences, Imperial College of Medicine, Norfolk Place, GB–London W2 1PG (UK) Tel. +44 171 594 3764, Fax. 44 171402 6861 Prof. Dr. K.W. Kühnel, Institut für Anatomie, MUL, Ratzeburger Allee 160, D–23438 Lübeck (Germany) Tel. +49 451 500 4030/4031/4001, Fax +49 451-500434, E-Mail kuehnel@anat.mu.luebeck.de
31.7.–3.8.2000 **Philadelphia** USA	Microscopy & Microanalysis 2000	**Information:** Mary Rebedeau, Fax +1 708 361 6166, E-Mail msa@tradeshownet.com http://www.msa.microscopy.com
10.8.2000 **Leuven** Belgium	Annual Meeting of the European Tissue Repair Society	**Information:** Mieke Flour, Tel. +32 16 337 870, Fax +32 16 337 951
3.9.–8.9.2000 **York** UK	11th International Congress of Histochemistry and Cytochemistry	**Information:** Prof. Dr. K.W. Schmid, Pathologisches Institut, Universität Münster, Domagkstrasse 17, D–48149 Münster (Germany) Tel. +49 251 83 55484, Fax +49 251 83 55 460, E-Mail schmihw@uni-muenster.de
9.9.–13.9.2000 **Hoeven** The Netherlands	T.C.E. 2000 (Tissue and Cell Engineering 2000)	**Information:** Tel. +31 20 444 8661, Fax +31 20 444 8683 E-Mail eh.burger.ocb.acta@med.vu.nl
1.10.–4.10.2000 **Rochester, N.Y.** USA	14th Rochester Trophoblast Conference, Meeting of the International Federation of Placental Association 2000	**Information:** Dr. Richard K. Miller, Department of Obstetrics and Gynecology, University of Rochester, Medical Center, 601 Elmwood Ave., Rochester, N.Y. 14642-8668 (USA) E-Mail rkmiller@obgyn.rochester.edu

Cells
Tissues
Organs

Acknowledgement to Referees

The Editors-in-Chief extend their gratitude and appreciation to the following reviewers whose comments and criticisms have ensured the quality of the papers published in this journal.

Klaus Addicks, Köln
Edythe L.P. Anthony, Providence
John D. Aplin, Manchester
Holger Bade, Köln
Augustinus Bader, Hannover
Rudolf Balling, Oberschleissheim
Michael Balls, Ispra
G. Beattie, La Jolla
Guy Benian, Atlanta
Klaus U. Benner, München
Dirk Berens von Rautenfeld, Hannover
W.A. Beresford, Morgantown
Martin Bergmann, Giessen
B.K.B. Berkovitz, London
K.J. Betteridge, Guelph
Andrew A. Biewener, Bedford
D. Blottner, Berlin
Ann-Marie Bollen, Seattle
Axel Brehmer, Erlangen
Dirk Breitkreutz, Heidelberg
Robert M. Brenner, Beaverton
Rainer Breul, München
Frede Bro-Rasmussen, Holte
Jennifer Bruechner, Williamsburg
Robert C. Burghardt, College Station
Dario Carrasco, Atlanta
Gwen Childs, Galveston
M.C. Colbert, Cincinnati
Ada A. Cole, Chicago
Bruce Crawford, Vancouver
Paul Dechow, Dallas
H.-W. Denker, Essen
Hugh Devlin, Manchester
Janice Dominov, Charlestown
Daryl Domning, Washington
A. Donner, Düsseldorf
S.J. Downing, York
Samuel C. Dudley, Decatur
Joseph T. Eastman, Athens
F. Ehrensperger, Zürich
R. Einspanier, Weihenstephan
Peter Ekblom, Uppsala
Hans H. Epperlein, Dresden
John Fallon, Madison
Colleen G. Farmer, Salt Lake City
Hend Farza, Headington, Oxford
W. Feichtinger, Vienna
Bernd Fischer, Halle (Saale)
Franz Peter Fischer, Garching
John-Gunnar Forsberg, Lund
Wolf-Georg Forssmann, Hannover
Armin E. Friess, Bern
Jonas Frisen, Stockholm
Steve Fuller, London
Franz K. Fuss, Vienna
Giorgio Gabella, London
Richard Gardner, Oxford
Renato Gerli, Siena
Ian Gibbins, Adelaide
Heiko Graichler, München
Paul Greenwood, Armidale
Robert Gregor, Atlanta
Peter Gruss, Göttingen
S.K. Gupta, New Delhi
James Hanken, Boulder
Hans-Georg Hartwig, Düsseldorf
S.G. Hassink, Wilmington
Axel Haverich, Hannover
Jerry L. Hedrick, Davis
Damon C. Herbert, San Antonio
Rainer Herken, Göttingen
John W. Hermanson, Ithaca
Susan W. Herring, Seattle
Andreas Herrler, Aachen
Anthony D. Ho, Heidelberg
Elisabeth Hooghe-Peters, Brussels
Curtiss D. Hunt, Grand Forks
William Hutton, Atlanta
Peter Hurst, Dunedin
Poul Hyttel, Frederiksberg
Taisen Iguchi, Yokohama
Yasuhiro Iwao, Yamaguchi
Peter Jecker, Aachen
K. Jewgenow, Berlin
Thomas Jungi, Bern
Najib Karim, Hannover
Tom Kennedy, London
Christoph Kirchner, Marburg
J. Kirsch, Ulm
Heikki A. Koistinen, Helsinki
Ron Kooijman, Brussels
T. Kozielec, Szczecin
Petra Köpf-Maier, Berlin
Wolfgang Kummer, Giessen
Thomas C. Lakars, Chicago
A. Lametschwandtner, Salzburg
Rudolf Leiser, Giessen
Csaba Leranth, New Haven
Ella S. Lindenbaum, Haifa
Douglas Luke, Kuala Lumpur
Kenneth T. MacLeod, London
Paul Maderson, Brooklyn
Wolfgang Maier, Tübingen
Ralph Marcucio, Ithaca
Robert Mayr, Wien
Siegfried Mense, Heidelberg
Karl Messlinger, Erlangen
Tony Metcalfe, Manchester
K. Militzer, Essen
Stefan Milz, Munich
D. Mitrovic, Paris
Ingrid Moll, Hamburg
Kazuo Momma, Tokyo
M. Montag, Bonn
Robert Moore, Freiburg
Joyce Morris-Wiman, Gainesville
John J. Mullins, Edinburgh
Magdalena Müller-Gerbl, München
W.L. Neuhuber, Erlangen
Peter Noakes, St. Lucia, Brisbane
Gerd E.K. Novotny, Düsseldorf
Frank Nürnberger, Frankfurt
Regis O'Keefe, Rochester
Andreas Oksche, Giessen
A.J. Olah, Bern 9
Timo Otonkoski, Helsinki
Reiner Peters, Münster
Françoise Pons, Montpellier
John Porter, Cleveland
Ralf Pörtner, Hamburg
Reinhard Putz, München
Enrico Reale, Hannover
R. Reimann, Graz
Russel J. Reiter, San Antonio
Marilyn B. Renfree, Parkville

KARGER
Fax + 41 61 306 12 34
E-Mail karger@karger.ch
www.karger.com

Accessible online at:
www.karger.com/journals/cto

C.L. Rieder, Albany
Paul Romert, Copenhagen
Roland R. Roy, Los Angeles
Roger H. Sawyer, Columbia
K. Schilling, Bonn
H. Schröder, Köln
Henk-Jan Schuurman, Basel
Kurt Schwenk, Storrs Mansfield
Ravindra M. Shah, Vancouver
Peter Sheridan, San Antonio
Charmaine J. Simeonovic, Canberra
Fred Sinowatz, München
Michael Sittinger, Berlin
Pieter Johannes Slootweg, Utrecht
Joseph Spadaro, Syracuse
Katharina Spanel-Borowski, Leipzig
Dan Spengler, Nashville
Udo M. Spornitz, Basel
William Stauber, Morgantown
Sharon Swartz, Providence
Kalman Szabo, Homburg
Clifford Tagoe, Accra
Julian R. Thorpe, Brighton
James Tidball, Los Angeles
Edda Töpfer-Petersen, Hannover
Norbert Ulfig, Rostock
Graciela Unguez, Austin
Warren Walker, Ossipee
Janna Waltimo, Helsinki
Richard Wassersug, Halifax
Andreas H. Weiglein, Graz
Ulrich Welsch, München
Frederick H. Wezeman, Maywood
Caroline Wigley, London
Hartmut Witte, Jena
Werner Wittkowski, Münster
John Wood, Atlanta
Karl Zilles, Düsseldorf

Cells Tissues Organs

Author Index Vol. 165, No. 3–4, 1999

Subject Index Vol. 165, No. 3–4, 1999

KARGER
Fax +41 61 306 12 34
E-Mail karger@karger.ch
www.karger.com

Accessible online at:
www.karger.com/journals/cto

Cells Tissues Organs

Author Index Vol. 165, 1999

KARGER
Fax + 41 61 306 12 34
E-Mail karger@karger.ch
www.karger.com

Accessible online at:
www.karger.com/journals/cto

Subject Index Vol. 165, 1999

Cells
Tissues
Organs

KARGER

Fax + 41 61 306 12 34
E-Mail karger@karger.ch
www.karger.com

Accessible online at:
www.karger.com/journals/cto

Vol. 165, 1999

Cells Tissues Organs

in vivo, in vitro

Founded 1945 as 'Acta Anatomica' by R. Chambers, G. Glimstedt, T. Peterfi and G. Wolf-Heidegger
Continued 1962–1974 by E.A. Boyden, 1955–1980 by A. Delmas, 1972–1980 by F. Walberg, 1945–1980 by G. Wolf-Heidegger, 1981–1988 by R. O'Rahilly, Davis, Calif., 1989–1990 by G.E. Goslow, Jr., Providence, R.I., 1981–1992 by W. Lierse, Hamburg

KARGER

Basel · Freiburg · Paris · London · New York ·
New Delhi · Bangkok · Singapore · Tokyo · Sydney

S. Karger
Medical and Scientific Publishers
Basel · Freiburg · Paris · London
New York · New Delhi · Bangkok
Singapore · Tokyo · Sydney

Drug Dosage
The authors and the publisher have exerted every effort to ensure that drug selection and dosage set forth in this text are in accord with current recommendations and practice at the time of publication. However, in view of ongoing research, changes in government regulations, and the constant flow of information relating to drug therapy and drug reactions, the reader is urged to check the package insert for each drug for any change in indications and dosage and for added warnings and precautions. This is particularly important when the recommended agent is a new and/or infrequently employed drug.

P.O. Box, CH–4009 Basel (Switzerland)
Printed in Switzerland on acid-free paper by
Reinhardt Druck, Basel

KARGER

Fax + 41 61 306 12 34
E-Mail karger@karger.ch
www.karger.com

Contents Vol. 165, 1999

Cells
Tissues
Organs

No. 1

Original Papers

No. 2

Original Papers

KARGER

Fax +41 61 306 12 34
E-Mail karger@karger.ch
www.karger.com

Access to full text and tables of contents, including tentative ones for forthcoming issues:
www.karger.com/journals/cto/cto_bk.htm

No. 3–4